诗意的功能主义

——德国格瓦斯·昆·昆建筑师事务所专辑

李保峰 译

中国建筑工业出版社

当代歌剧与音乐中心，柏林，2001年
CENTRE FOR CONTEMPORARY OPERA AND MUSIC, BERLIN, 2001

1   20

日本2005年世界博览会德国展馆，日本爱知县，2004年
GERMAN PAVILION AT THE EXPO 2005, AICHI/JAPAN, 2004

2   24

奥林匹克建筑综合体 "体育的动感"，莱比锡，2003年
OLYMPIC BUILDING COMPOUND "SPORT MOVES", LEIPZIG, 2003

3   28

医疗科技博物馆，埃尔兰根，2003年
MUSEUM OF MEDICAL TECHNOLOGY, ERLANGEN, 2003

4   32

THYSSENKRUPP学院，埃森，2006年
THYSSENKRUPP ACADEMY, ESSEN, 2006

5   36

漂浮之家，马尔马拉海岸，2003年
FLOATING HOME, MARMARA COAST, 2003

6   40

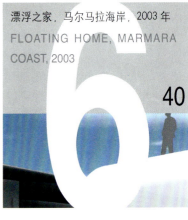

地标建筑与水下酒店，青岛，2007年
LANDMARKBUILDING WITH UNDER WATER HOTEL, QINGDAO 2007

7   44

帕莱斯垂直"花样住宅"，波兰，华沙，2004年
PALAIS VERTICAL "BLOSSOM HOUSE", WARSAW/POLAND, 2004

8   48

完全购物中心，埃森，2005年
TOTAL SHOPPING, ESSEN, 2005

9   52

奥迪公司变速箱及排放研究中心，英戈尔施塔特，2005年
AUDI LTD GEARBOX AND EMISSIONS CENTRE, INGOLSTADT, 2005

10   56

中央火车站新站，慕尼黑，2006 年
NEW CENTRAL RAILWAY
STATION, MUNICH, 2006

60

白南准博物馆，韩国京畿道，2003 年
NAM JUNE PAIK MUSEUM,
KYONGGI/SOUTH KOREA, 2003

66

商业和贸易展览中心，比勒费尔德，2000 年
COMMERCIAL AND TRADE FAIR
CENTRE, BIELEFELD, 2000

70

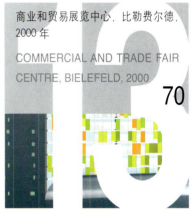

体育竞技场，柏林，2005 年
SPORTS ARENA, BERLIN, 2005

74

悠闲巡游，2006 年
EASY CRUISE, 2006

78

富士通机器人总部，斯图加特，2005 年
HEADQUARTERS FOR FANUC ROBOTICS, STUTTGART, 2005
84

国际航空和空间技术中心，柏林，2003 年
INTERNATIONAL CENTRE FOR AVIATION AND SPACE TECHNOLOGY, BERLIN, 2003
90

马克西米利安和宫廷马厩广场的州歌剧院，慕尼黑，2003 年
STATE OPERA AND MAXIMI-LIANHÖFE ON MARSTALLPLATZ, MUNICH, 2003
94

M 星球—贝塔斯曼展馆，2000 年世博会，汉诺威，2000 年
PLANET M – BERTELSMANN PAVILION EXPO 2000, HANOVER, 2000
100

德利佳华办公塔楼，法兰克福／美因河畔，2003 年
OFFICE TOWER FOR DRESDNER KLEINWORT WASSERSTEIN, FRANKFURT/MAIN, 2003
104

联网天然气股份公司总部，莱比锡，1997 年
HEADQUARTERS FOR VERBUN-DNETZ GAS AG, LEIPZIG, 1997
108

戴姆勒克莱斯勒航空大楼，柏林，2000 年
DAIMLERCHRYSLER AERO-SPACE, BERLIN, 2000
114

新火车站，东柏林，2000 年
NEW RAILWAY STATION, BERLIN EAST, 2000
118

音乐会和娱乐建筑，柏林，2007 年
CONCERT AND ENTERTAIN-MENT BUILDING ADMIRALSPA-LAST, BERLIN, 2007
122

"SOPHIE–GIPS–HÖFE"，柏林，1997 年
"SOPHIE-GIPS-HÖFE", BERLIN, 1997
126

前言 PREFACE **7**	新作 + 精选 UNRELEASED > RELEASED **18**
格瓦斯 · 昆 · 昆访谈 INTERVIEW WITH GEORG GEWERS, SWANTJE KÜHN AND OLIVER KÜHN CLAUS KÄPPLINGER **8**	传记 BIOGRAPHIES **134**

左起：乔治·格瓦斯
奥利弗·昆
斯万提·昆

摄影：Udo Hesse

前言

如果成就可以创造出对新梦想的渴望,那么经验能走向新的起点。这种想法促成我们出版这本作品集。这些呈现的作品凝聚着我们的心血,我们十分珍惜。

15 + 10反映了竞赛入围方案、建筑概念设计以及建筑项目。这些设计(尤其是在某些新的探索领域中)极大地挑战了我们的创新潜质,拓展了先前的设计语汇,并激发了我们不断寻求解决问题的最佳方案的热情。

这是一种十分个性化的汇编。我们当然希望所有的项目都得以实施。但是,我们并不把实施项目的多少作为评价我们成绩的惟一标准。

我们寻求创造性的挑战。我们希望在纷繁复杂的要求下寻求完美的解决方案,从而赋予建筑形象。我们为之努力,我们有不灭的热情。

PREFACE

Experience inspires initiative if one's own achievements create a longing for new dreams. This thought prompted us to publish a book dedicated to those things that make our work worth while and valuable to us.

15+10 is a reflection on competition entries, architectural concepts and projects which, to a high degree, have challenged our creative potential, broadened our formal language and promoted our passionate undertaking of finding the best possible solution, especially if this process encouraged the exploration of new areas.

This compilation is a very personal choice. Naturally, we would have loved to have realised all of these projects. However, we do not measure our achievements solely by the number of projects actually realised.

It is the creative challenge that we seek, the wish to find the perfect solution that inspires us to fill our architecture with form, content and spirit against the background of various demands. This is what we are striving for. With unbroken enthusiasm.

访谈

你们这本出版物的标题是"Unreleased—Released 15 + 10"。那么请问是什么促使你编辑这本以15个未发表的方案和10个发表过的项目来代表你们设计团体的书？

乔治·格瓦斯：

设计是我们工作的基础。但是并非我们所有的设计方案一定都能进入到最后的成果阶段。于是我们利用"15 + 10"这样一个很好的机会介绍那些未被介绍的设计方案。

奥利弗·昆：

建筑一般是建造起来并且是要被看到的。如果设计的一座建筑没有最后建造完成，这样的设计就束之高阁。这并不意味着它没有价值，许多设计被证明太优秀以致于不能在当前被付诸实施。我们经常从一开始就被抛弃的方案中重新拾起这些设计想法。因此，我们决定出版"15 + 10"以展示我们建筑的实验性以及建立在其实验性上的设计过程。

很明显，"创新"已经成为你们创作的中心。你们在新建筑和想像中非常传统的建筑两方面一次又一次成功地飞出了解决方案。"创新"对于你们来说到底意味着什么呢？

斯万提·昆：

到底什么是真正的创新？是寻找另一个更疯狂的形式？不！所谓创新就是在任何既定的设计过程中超越常规以寻找正确的解决方案。在商业开展中，方盒子的造型经常被认为是惟一可行的解。不，并非在任何时候都是方盒子，而是与设计过程相适应的，反映公司识别性的外形。创新就是寻求有效解答、超越常规之路。

乔治·格瓦斯：

创新也可理解为"发现"和"创造"，在发现之旅中寻找新的空间和组合新的建筑材料。2005年日本世博会爱知展馆就是一个很好的例子，它的空间是以前从未被体验过的。创新也被看作是一种在技术上和构造上受欢迎的挑战。举例来说，迪拜之塔是一个和SBP工程事务所（Schlaich, Bergermann与合作者）一起合作的项目。通过推进静力学的极限和应用新的构造方式，迪拜之塔达到了至少440米的高度。

奥利弗·昆：

创新是非常多层面的，它能运用到任何地方。创新能够影响材料的使用、形态和设计。它能够带来令人惊讶的智能功能统一体。富士通自动数控公司机器人总部项目展现了对智能商业再设计的创新。关于设计过程我们知道什么？在何种程度反映建筑的真实形态？建筑如何根据商业需求在尺度上进行扩大或者缩小？当前传统的"形态化"和"蜻蜓点水"对我确实不再有帮助。然而，一座在尺度上可变的建筑，运用适宜的结构技术支撑能够缩放到正确尺度，这种交互式的进展是一种真正的创新过程。它不是一种实验式的造型过程，而是一种引导新解的分析方法。

技术的进步并非总是与当今社会的发展同步。对于这些不同的发展步调以及随之而来的问题，你们是怎样处理的？建筑师经常抱怨被强加给创新太多的社会约束。

奥利弗·昆：

不同于许多其他建筑师同行，我们并不抱怨现实过程中的很多约束。正好相反，我们相信由不同的当权者以及客户的要求的诸多规范为创新提供了一片丰富和肥沃的土壤。在那些大众真心出谋划策的国家会有创新的建筑，比如在瑞士，在荷兰这样一个高度多元化的市民社会国家也会找到。英国是另一个例子,而德国正在迎头赶上。但是我们没有在美国、俄罗斯和中国找到。创新的推动力只能产生于一种相互关联的、许多利益群体共同承担享有的多元化体系中,而不是在一个"诸事走开",单一因果关系的、格律诗般的、缺乏民众关心所支配的体系中。

包括建筑师在内的很多人观察到当前的变化发生如此之快,以至于个体无法承受这种变化带来的压力。他们认为通过创造对过去时代产生回忆的空间以减缓这种变化就是建筑师的任务。你们是不是觉得建筑师应该对这种飞速的变化做出某种补偿?

乔治·格瓦斯:

同十年前相比,我们的确面临着一个带着更多复杂问题的复杂世界。建筑也变得非常多元化,不仅仅需要高水平的工作团队,而且需要新型的交流构架。这里不需要任何怀旧形式主义的东西。

斯万提·昆:

我确实不愿推荐通过建筑的方式试图减缓城市发展的脚步。复杂性既然已经存在,那么我们所有人就日复一日使用全部的技术解决它。然而,我相信我们需要简化复杂性,以便于我们深入认识问题,更好地解决问题。因此这个问题的答案不是倒退到所谓"昔日好时光"中。我们需要一种进步的方法,承认当今东西的优点。然而,我们希望能使它们更为简单化和更容易理解。

在建筑师工作中相互沟通这一点上,你们似乎在追求着更强的竞争力。为了把握当代的复杂性,你们以学科交叉团队的方式工作。

奥利弗·昆:

学科交叉可以催开灵感之花。这种交叉是我们一个重要的着力点。关键是各专业配合!例如,在巴洛克艺术中,专业配合是无所不在的。在巴洛克教堂的建造中,没有建筑师会担心后来的画家和雕塑家出现在室内装修中,也没有画家和雕塑家会担心后来的风琴弹奏者用音乐修饰教堂。正相反,来自不同学科背景的人们团结起来,协同工作并产生对每个其他专家的作品共鸣般的信任。这就是我们今天所需要面对的。建筑并不是单独的一个专家的事情,我们探究团队设计的真正本质。

在讨论柏林当代歌剧与音乐中心的设计时,我们清楚地认识到我们需要将作曲家、音响工程师、乐队指挥和音乐家纳入到设计过程中。我们也需要咨询筹款者、基金会、文化部门和编辑。最后的结果总是更完善,并且比起我们曾经的单独行动,我们都感觉这样的团队努力要好很多。

你们几次强调,从一个项目的初期阶段开始就纳入各种艺术家。事实上,许多你们的同行故意避免与艺术家协商而自己去创造一件艺术作品。请问你们这样的态度背后的动机是什么?

奥利弗·昆:

这样的建筑师确实有很大的错觉!确实有建筑师不允许他们的住户悬挂任何私人照片,担心会把他们的"艺术作品"毁坏。他们确实没有理解建筑的本质。建筑是为他人建造的。是的,我们是明白某种设计想法的那类人,但是设计的着眼点在于他人和公众。我们行业协会名誉的降低应该归咎于这种过分的傲慢。如果建筑不能充分证明允许他人的介入,那么它无论如何不太可能走到最后。你只需要从头到尾地思考整个过程。与艺术家合作,我们总是寻找到能够发展强烈而独立的想法。宫廷马厩广场就是这样空间的一个很好案例。奥拉维尔·埃利亚松已设法正式改善它并为其增加协作的价值。

乔治·格瓦斯:

我们的许多项目一开始就和艺术家们合作设计。与艺术家早期的对话是一面有趣的镜子,将我们的想法反射给我们自己。现在职业所及的范围未被清晰地定义；所以一座理想的建筑应该是所有艺术的产品。

如果建筑师和画家、雕塑家以及当今的图形设计师、灯光艺术设计师以及来自许多其他学科的人合作我不会感到困扰。

斯万提·昆：

另一方面，社会中的设计作品和情感价值的联系越来越多，甚至建筑上的情感层面也变得更为重要。我们需要触发情感的层面，并与超越建筑物边界的影响相联系。艺术家们以不同的角度观察事物，也是我们将他们纳入项目设计团队的另外一个原因。我们建筑师的任务是全局性地推进设计过程，而艺术家们人物单纯，特立独行。

你们十分开明，允许其他人参与你们的设计。你们似乎也对后来的住户怎样使用建筑十分放心。

斯万提·昆：

音乐能够打开和关掉。你也会选择去参加一个艺术展览会或者错过它。艺术几乎总是互动的，然而建筑未必这样。你不能简单地关掉建筑，而是不断地在触摸它。一栋受欢迎的建筑能让你进入并与之交流。然而，那些不能被感触的建筑物根本就不能引人入胜。建筑必须适于交流。我们设计的建筑欢迎住户入住，并且按照他们喜欢的方式进行改变。

奥利弗·昆：

一旦第三方涉入，一个人对事物的最初想法就会得到重新阐释。我们常谈论波尔坦斯基，他就是一个研究这个现象的艺术家。他将空盒子放在房间里，鼓励来访者将自己的故事作为艺术家已有叙述的延续去填充这些盒子。重新解释就这样作用于我们的建筑。它创造了一个能被每个个体重新思考的世界。我认为正是这样赋予了我们建筑的乐趣。建筑能被个别解释了，这种立场与所谓的理性主义者形成鲜明对比，他们声称"建筑不应留有解释的余地。我们的观点坚如磐石，那就是尽头"。

你们书中的设计案例并不遵循一种固定的设计模式、形态模式或者色彩语言。我们看到不同的设计有着不容否认的个性品质，这些个性品质总是被赋予多样化的涵义。请问这是如何产生的？

斯万提·昆：

我们三人有共同的背景——我们都是在英国获得我们的专业技能。进一步分析，我们所有的设计都有一个普通且十分苛刻的结构图式，合乎逻辑的原则指导着所有设计。接着就是我们的不同之处。乔治是一位雕塑家，我在美国学习美术，奥利弗则在圣迦南大学就读管理学。于是我们都从另外的领域吸取灵感。并不只有建筑，我们心中另外的世界丰富着我们。

奥利弗·昆：

我同意。尽管我倾向于分析方法，但还是会对那些唤醒我或者让我有某种暗示的事物作出非常直觉的反应。自从我们成为三人团队以来，从不同的领域吸取灵感对于我们所有的人都非常重要。我们最初通过大量图像界定设计项目，然后发展出共同工作的导则。有时候从任务而来的图就已经清楚了，有时候从应用的技术、印象或者我们为设计制定的任务说明书等而来。在任何案例当中，图像必须保证每个人都能很好地理解消化。结果从各个不同设计发展出毫不妥协的设计原则。

从你们的设计案例和众多的讲演中可以发现，你们认为建筑必须大于其他各部分之和。即使是一个工业预制建筑应该富有可感觉的个性和某种程度上的知觉。但是协

同设计理念并不正是要求相似的建筑表达，允许置换到世界任何地方吗？

乔治·格瓦斯：

协同设计不是给出"标准答案"。它更倾向于针对不同项目寻找特别方法。毕竟，我们是在特殊的场所为有希望的、特别的客户建造。

在绘制第一张满足客户和场所要求的图纸之前，我们考虑自己对功能的"消化作用"以努力了解到尽可能多的手头上的设计任务。

这个过程对于我们非常重要。通过不断结合最初的设计任务和现实条件，我们才能提出高质量的解决方案。当今的客户和住户都在寻找较之以前更多的个性化的接触，同样适用于工业构筑物。平庸的方盒子不会使任何人愉悦，即使这样的建筑是新的，它们也并不动人。我们的城市确实需要个性和世故，战后德国的城市并非典型的美的范例。

奥利弗·昆：

我相信与建筑及其所在地相关的情感联系是很重要的。老房子应有自己的名字！在今天这个数字化的年代，我们对于个体情感局限有着更为强烈的渴望，而这种情感与具体的个体和场所紧密结合。这样，建筑重新获得那些久违的独特性和真实性。对个性的渴望既不能由新"国际式"实现，也不能指望某些建筑师设计的、能够在世界任何地方出现的"标签式建筑"。我们一再强调"某个房子只能属于某个地方，在这样的环境中它是独一无二的"，这很重要。

许多建筑师谈到他们的建筑所拥有的某种个性。但是个性是一个纯粹个人的术语，它是个体的、主观的、自圆其说的。建筑没有主观性，它只能为社会和沟通提供空间，这个空间也许满足一个人对于个性的需求。在最好的案例策划中，住户易于识别建筑。在这个意义上，你们如何看待你们的建筑？

奥利弗·昆：

建筑个性的主题贯穿于公司员工的互动式沟通的始终。举例来说，"居家的感觉"和"取得联系"的主题在建筑中具有重要和基本的意义。在我们的设计中，富士通自动数控公司项目在平台上设置一处日本式庭园，以鼓励他们全体职员轻松交流。"Sophie—Gips—Hofe"项目提供了一些不同类型的人们可以约会的公共场所。宫廷马厩广场上的外部公共空间，对于我们来说甚至比建筑本身更为重要，用于放松和交流的中庭和内部全景在法兰克福项目中得以体现。人们可以在这些地方交流相处。建筑的可识别性独立产生于社会互动式交流，人们的聚会和与所在地相结合的个人体验。慕尼黑宫廷马厩广场得到重建，最后的结果公布于众，人们拍手称赞。有位年长的女士说她又成为了在炸弹摧毁慕尼黑之前的那个小女孩。建筑师具有催生如此情感体验的力量。这些主要因素决定了人们接受或拒绝某个地方，这是一个场所的灵魂，它赋予场所超越全部的商业功能的价值，这就是为什么建筑在我们的社会生活中如此重要的原因。

乍看之下，在参与全球化设计和地方性设计的建筑之间，在工业化、合理优化和建构独特个性之间存在矛盾。怎样定义当代建筑个性呢？

乔治·格瓦斯：

工业化预制并不是必然意味着更少的个性。由工业化预制构件建造的一些建筑诚然是令人厌烦的，但是发展一座有着特殊品质的、在个性化范畴内的建筑，人们希望有着个性的建筑更多的是一个拥有正确精神的、关于知性地掌握现代结构方法的问题。我们非常高兴地将20世纪60年代的"国际式"和80年代的"后现代主义"抛诸脑后。今天，更为个性化，赋予建筑某些特质成为建筑师的目标。通过一座建筑可以给人感观效果、不同特性、感染力和情感的凝聚。

斯万提·昆：

我们需要清楚地把握场所的确切性质。当然，为贝塔斯曼这样一位全球性角色建造展馆不必限于特定的场所。它的建造地点也不是真实的场地，这个展馆不是公司的家，而是一个公共陈列窗口用以展示公司的个性与形象。在某种意义上，这个展馆本身已成为一个场所；这个独立的地标处于汉诺威郊区，它的本质和联系只不过在反映特定的企业文化上是明显的。

然而，"Sophie-Gips-Hofe"庭院则位于一个非常具体的地方，这是个被遗弃的、讨人喜欢的地方，这个地方还一度成为小镇的一部分。我在这里坐了一天时间来观察所有的旅行者和当地居民，我自己想："我在这里坐下来看那些真正喜欢小镇的人，过了一会儿他们就占据了这个地方，似乎这里就是他们的家"。这是一种难以置信的愉悦感。

奥利弗·昆：

我特别喜欢这种区别。什么是一个设计项目的家园？在那里建造建筑，它是一个企业、企业形象或者它是建造建筑的真正场所。对我而言，"全球的"和"个体"之间或"个性"和"工业的"之间没有矛盾。大规模的装配、加工的数字化生产线和全球联盟很久以前就成为任何产品生产过程的伴随因素。它们限定了我们的时间并决定了我们所使用的成套结构工具，它们也创造出新的选择。让我们看看在华沙的"垂直的宫殿"，目前那里有一支跨国的设计团队和横贯大陆的金融家。但是，仍有一个十分明确的、拥有20世纪70年代建成的居住街区的场所。这个场所是独特的，但实际上是城镇的老区。尽管对这样一个建筑通常所有的高科技是十分必要的，结果成为一种展示波兰和华沙进入欧盟的特殊开发案。虽然不同的国际利益团体参与到这个项目中来，但这个建筑并不能够在世界其他任何地方得以建成。正如个体的环境一样，在全球化中我们需要更为关注可识别性、天才的思路和项目的个性。这正是将个体的建筑持续到未来的机会。

克劳斯·开普林格 采访
格瓦斯·昆·昆

INTERVIEW

The title of your publication is "Unreleased – released 15+10". What prompted you to compile a book of 15 unreleased and 10 released projects representing the work of your partnership?

Georg Gewers:
Design is an essential part of our work. However, not all the projects we design necessarily enter into the production process. "15+10" is a great opportunity for us to present those projects that would normally not be shown.

Oliver Kühn:
Architecture is there to be built and to be seen. If a building is designed but doesn't end up being built its design disappears in a drawer. That doesn't make it valueless, though; many ideas prove just too good to be carried out at the present time. We have often resurrected ideas contained in projects that had initially been discarded. Consequently we have decided to publish "15+10" to show the experimental aspects of our architecture as well as processes that are based on one another.

Innovation has obviously always been a central aspect of your work. You have succeeded time and again in producing architectural solutions to both new and supposedly very conventional constructions. What does innovation mean to you?

Swantje Kühn:
What indeed is innovation? Finding yet another wild shape? No! Basically innovation is about finding the right solution for any given production process beyond convention. In commercial development box shapes have often been regarded as the only viable solution. No, it doesn't necessarily have to be a box all the time but rather a shell suitable to the production process, reflecting the company's identity. Innovation is the path of finding efficient solutions beyond conventional methods.

Georg Gewers:
Innovation is also about "discovering" and "inventing", about finding new spaces and combining new building materials that take you on a tour of discovery. The AICHI Pavilion for the EXPO 2005 in Japan is a fine example of experiencing space as never experienced before. Innovation also poses a welcome challenge in terms of technology and construction. Take, for example, the Dubai Tower, a project shared with "Schlaich Bergermann + Partner" Engineers. By pushing the limits of statics and by applying new construction methods the tower reaches a height of no less than 440 metres.

Oliver Kühn:
Innovation is extremely multi-layered and can be applied anywhere. Innovation can affect materials used, form or design. It can also bring about a surprisingly intelligent composition of functions which develop into new synergies. The FANUC project displays innovation as an intelligent business reengineering. What do we know about the production process? To what extent is it reflected in the actual shape of the building? How can the building be scaled up or down in size according to business needs? Today's conventional 'morphing' or 'blobbing' doesn't really do it for me. However, the interactive development of a building variable in size, one that can be zoomed into the right scale with the help of a suitable supporting structure is a truly innovative process. It's not an experimental shaping process, but an analytical approach which leads to new solutions.

Progress in technology isn't always synchronous to the developments in our society today. How do you deal with the different pace of those developments and the problems resulting from it? Architects often complain about the many social restrictions imposed on innovation.

Oliver Kühn:
In contrast to many of our colleagues, we don't complain about the many restrictions in the realisation process. Quite the opposite, we believe that the extensive regulations imposed by the various authorities as well as the clients' demands provide a very rich and complex soil for innovation. Innovative architecture is found in countries where the opinion of the public really counts, such as in Switzerland. It's also to be found in Holland, a highly pluralistic civil society. England is another example and Germany is just beginning to catch on. However, we do not see it in the USA, Russia or China. The impulse for innovation can only be generated in an interconnected pluralistic system with a number of interest groups partaking rather than in a system where "anything goes", where mono-causal, metrically oriented structures dominate without any public input to speak of.

Many people, including architects, observe that nowadays change takes place at such a high pace that it puts an unbearable strain on the individual and they reckon that it is architecture's task to slow down this pace by creating spaces reminiscent of a past era. Do you feel architecture should somehow compensate for the rapid pace of change?

Georg Gewers:
Compared to 10 years ago we certainly are facing a much more complex world with much more complex procedures. Architecture, too, has become extremely multi-dimensional and not only requires team-work of the highest standard but also new structures of communication. There is no need for any kind of retro-formalism.

Swantje Kühn:
I really wouldn't recommend trying to slow down a city's pace by means of architecture. Complexity already exists and we all support it with all the technology in use on a day to day basis. However, I believe we need to simplify it, so that we can deepen our understanding and make better use of it. Therefore the answer to the problem cannot be to step back in time pretending that "things were better in the old days". We need a progressive approach which acknowledges the high quality of many things today. However, we want to help to make them even simpler and more comprehensible.

You seem to be pursuing a higher degree of competence as architects on a communicative level through your work. In order to master today's complexity you work as an interdisciplinary team of experts.

Oliver Kühn:
Inspiration blossoms where disciplines overlap. This kind of overlap is an important issue to us. Keyword: synergies! In baroque art, for example, synergies are omnipresent. No baroque church would have been built had the architect been afraid of the painter or the sculptor who would later on decorate the interior. And neither painter nor sculptor would have been afraid of the organist who would later fill the church with music. Quite to the contrary, people from different disciplines were drawn together in order to form synergies and synaesthesia trusting in each other's expertise. This is exactly what we should be aiming for today. Architecture is not solely a specialist's concern. We ought to explore the very essence of the project together as a team.
When discussing the design for a contemporary opera house in Berlin it was clear to us that we needed to involve composers, sound engineers, opera directors and musicians in the process. We also needed to consult fundraisers, foundations, cultural movements and editors. The final result is always much better and we all feel better about such a team effort than we would have done had we gone "solo".

You have repeatedly made a point of involving artists right from the very early stages of a project. What is your motivation behind this attitude, considering many of your colleagues deliberately avoid consulting artists and proceed to create a work of art by themselves?

Oliver Kühn:
Such architects really have delusions of grandeur! There are even colleagues who won't allow their tenants to put up any private pictures in fear of having their "work of art" ruined. They have surely failed to comprehend the essence of architecture. Architecture is produced for other people. Yes, we are the ones realising an idea, but the focus is on other people and on the public.
Our guild's dwindling reputation is due to just this kind of arrogance. If architecture doesn't prove robust enough to allow for the intervention of others, it is unlikely to last anyway. You only need to think the whole process through from beginning to end. Together with artists we have always sought to create spaces that can develop into something truly strong and independent. The Marstallplatz is a fine example of such a space. Olafur Eliasson has managed to both improve it formally and to add synergetic value to it.

Georg Gewers:
Many of our projects are initially designed in cooperation with artists. An early dialogue with artists is an interesting mirror in which our own ideas can be reflected back to us. Nowadays areas of competence aren't defined all that clearly anyway; a building should ideally be a product of all the arts. It really doesn't bother me if architects

collaborate with painters and sculptors and nowadays graphic designers, lighting artists and people from many other disciplines, too.

Swantje Kühn:
There is yet another aspect to it. In a society where every product is increasingly associated with an emotional value, the emotional layers on architecture become even more important. We need layers that trigger feelings and associations that extend the impact of a building beyond its metric boundaries. Artists see things differently and that's another reason why we involve them in our projects. They are exclusively content-orientated and independent, whereas our task as architects is to guide and manage the process as a whole.

You are surprisingly open-minded towards letting other people contribute to your projects. You also seem very relaxed about how your tenants actually intend to use the building later on.

Swantje Kühn:
Music can be switched on and off. You may choose to go to an art exhibition or you may choose to give it a miss. The arts are almost always interactive whereas architecture isn't. You can't simply switch it off, you are constantly in touch with it. A well-received piece of architecture allows you to step right inside and to interact with it. However, buildings that are objects which mustn't be touched don't go down well at all. Architecture must be fit for dialogue. The tenants in our buildings are welcome to settle in and make changes if they like.

Oliver Kühn:
As soon as a third party gets involved, one's own vision of things gets reinterpreted. The other day we talked about Boltanski, an artist who works with exactly this phenomenon. He places empty boxes in a room encouraging the visitor to fill these with their individual stories as a personal continuation of the artist's own narrative. That's how it is with our architecture. It creates worlds which can be rethought by everybody individually. I think that's exactly what gives us pleasure in architecture, architecture which can be interpreted individually. This stands in striking contrast to the so-called rationalists who claim, "Architecture should not leave room for any further interpretation. Our view is cast in stone and that's the end of it."

The projects in your book do not follow a fixed planning pattern, model form or colour coding. We see different designs each of which have an undeniable individual quality always endowed with multiple meaning. How does this come about?

Swantje Kühn:
The three of us do have one thing in common – we all acquired our skills in England. On closer examination there is a generic and absolutely stringent structural diagram common to all our designs, a logical principle governing all. And then there are our different personalities. Georg is a sculptor, I studied Fine Arts in the States and Oliver read Management at the University of St. Gallen. So we all draw inspiration from additional fields. It is not architecture alone. There are other worlds in us that enrich us.

Oliver Kühn:
I agree. Although I generally tend to an analytical approach I react quite intuitively to things that remind me or let me have an inkling of something. Drawing inspiration from different fields is very important to all of us, since we work as a trio. We initially define a project via a number of images. Then we develop a guideline which defines the direction we want to go together. Sometimes an image is already clear from the task on hand, sometimes it evolves from the technology used or from a vision or "mission statement" which we develop for the design. In any case, the image must always remain comprehensible to everybody involved. The result is a clear, uncompromising design guideline from which various highly differentiated designs are developed.

In your projects and your numerous lectures you adopt the position that more than ever before architecture must be more than the mere sum of its parts. Even an industrially pre-fabricated building ought to be of a tangible individual character and possess a certain degree of sensousness. But doesn't the very idea of Corporate Design demand a homogeneous architectural expression which will allow for a translocation to any place in the world?

Georg Gewers:
Corporate Design isn't about giving "standard answers". It's rather about finding unique solutions to different tasks. After all, we build on specific sites for, hopefully, specific clients. We consider ourselves functional "contextualists" striving to learn as much as

possible about the task on hand before drawing the first design tailored to both the client's and the site's needs.
This process is very important to us. By reconnecting to the initial task and situation we are able to come up with high quality solutions. Nowadays clients and tenants alike are looking for that individual touch more than ever. This also applies to industrial construction. Banal boxes won't make anyone happy. Such buildings are often unoccupied even if they are new. Our cities really could use a touch of individuality and sophistication. German post-war cities aren't exactly prime examples of beauty.

Oliver Kühn:
I believe the emotional attachment to a building and its location is very important indeed. Houses used to have names! Today, in the age of digitalisation, there is an even greater yearning for emotional bonding which is individual and locates us concretely. In this way architecture regains those long-lost qualities of uniqueness and authenticity. This longing for individuality cannot be fulfilled either by a new "International Style" or so-called "Signature Buildings" attributed to certain architects, which could stand anywhere in the world. It is important to be able to say once again, "This house could only be in this place, in this context – it is quite unique."

Many architects speak of a certain identity that their buildings possess. But identity is a purely personal term related to the individual, subjective process of finding self-reassurance. A building cannot be a subject, it may only offer space for socialising and communicating which may support a person's quest for identity. In the best case scenario the tenant will identify with the building. How do you see your architecture in this perspective?

Oliver Kühn:
The theme of identifying with a building through communicative interaction of a firm's employees, for example, the theme of "feeling at home" and "getting in contact" are of great and fundamental importance in our architecture. Take a look at our designs: The "FANUC"-project entails a Japanese garden on a platform encouraging their members of staff to socialise freely. The Sophie-Gips-Höfe project offers several different public places where people can meet. The external public spaces on Marstallplatz were almost more important to us than the buildings themselves. The atrium and the Inside Panoramas for relaxation and interaction offered at our Frankfurt project – these are all places where people can get together. Identification with a building is generated exclusively by social interactions, by meeting people and by individual experiences which bond one with that place. When the Marstallplatz in Munich was reconstructed and the result finally unveiled people applauded, and there was an old woman who said, she had once again become the little girl she had been before the bombs destroyed Munich. An architect does have the power to generate such emotional experiences. These vital factors determine whether people accept or reject a place. It is the soulfulness of a place which endows it with values beyond its general commercial function. This is what makes architecture so important to our social life.

At first glance there seems to be a discrepancy between being a global player on the one hand and creating architecture for a very specific place on the other, between industrial prefabrication, rational optimisation and buildings of a unique individual character. What defines individuality in architecture nowadays?

Georg Gewers:
Industrial prefabrication does not necessarily mean less individuality. Some buildings made of industrially prefabricated elements are admittedly very boring indeed. But developing a building with special qualities, with scope for individuality, one that people want to identify with is more a question of having the right spirit, about mastering modern construction methods intellectually. We gladly left behind that "International Style" of the 1960s and the "Post-Modernism" of the 1980s. Today, more individuality is the goal of architects, giving a building something unique. This is perceptible through a building's sensuous effect, well-differentiated features, charisma and an emotional density.

Swantje Kühn:
We need to be clear about the exact nature of the location. Of course, the pavilion for Bertelsmann, a global player, is not tied to a specific location. Its location is not a real place, the pavilion is not the company's home, but a public showcase displaying the company's identity and image. In some sense the pavilion itself has become the place; a solitary landmark on the outskirts of Hanover whose roots and connections are only apparent in the reflection of a specific corporate culture.
The Sophie-Gips courtyards, however, are situated at a very concrete place, a lost, enchanted place which had once again become a part of town. This is where I sat one day observing all those tourists and locals, thinking to myself, "Here I sit looking at people who really like this part of town. Just for a little while they occupy this place as if it were their own home." It was an unbelievably happy feeling.

Oliver Kühn:
I like this differentiation a lot. What is the home of a project? Is it an enterprise, a Corporate Identity or is it the real location where construction takes place?
To me there is no discrepancy between "global" and "individual" or between "individuality" and "industrial". Mass fabrication, digital lines of processing and global alliances have long ago become accompanying factors of any production process. They define our times and determine the construction kit with which we work. They also create new options. Let's take a look at the "Palais Vertical" in Warsaw: present were a transnational planning team and transcontinental financiers, yet there was a very definite site with residential blocks from the 1970s and the picturesque, but basically virtual old part of town. Despite all the usual high-tech necessary for such a construction the result was a specific development representing Poland's and Warsaw's entry into the EU. Although various international interest groups took part in the project, this house could not have been built anywhere else in the world. In the global no less than the individual context, we need to take greater care over examining the identity, genius loci and individuality of a project. And exactly here is where the chances for lasting individual architecture in the future lie.

Claus Käpplinger interviewing
Gewers Kühn + Kühn

UNRELEASED 新作 i5+10

当代歌剧与音乐中心，柏林，2001 年
形同乐器的建筑

CENTRE FOR CONTEMPORARY
OPERA AND MUSIC, BERLIN, 2001
A BUILDING JUST LIKE AN INSTRUMENT

当代音乐表演对场地需求的不断增长，产生了这个全新的概念"当代歌剧与音乐中心"。格瓦斯·昆·昆本能地将这一项目的选址定在柏林新的中心车站附近，这里将是连接东西欧的洲际列车的重要节点。格瓦斯·昆·昆每年都要将自己置身于新的挑战之中，这已不仅仅是竞赛的需要或客户的要求，而是自发探索一种更新的创作方式。

这个项目一开始就面临的主要问题是：新建的音乐厅或歌剧院仍旧按照老式风格的古典音乐的场地进行设计，几乎没有给当代音乐提供合适的场地。

It was the growing demand for specific performance venues of contemporary music that prompted this totally new concept for a "Centre for Contemporary Opera and Music". On their own initiative GKK developed a building to be located near Berlin's new Central Station, the future continental railway junction connecting eastern and western Europe. Every year GKK set themselves a new challenge outside the requirements of a competition or a client's demands in order to explore new methods for themselves.

The starting point for this project was the fact that even new concert halls and opera houses still follow the old-fashioned room design for classical music venues and that there are hardly any appropriate venues for contemporary music.

外观像一道被音箱围绕的风景
a façade like a landscape wrapped around a sound box

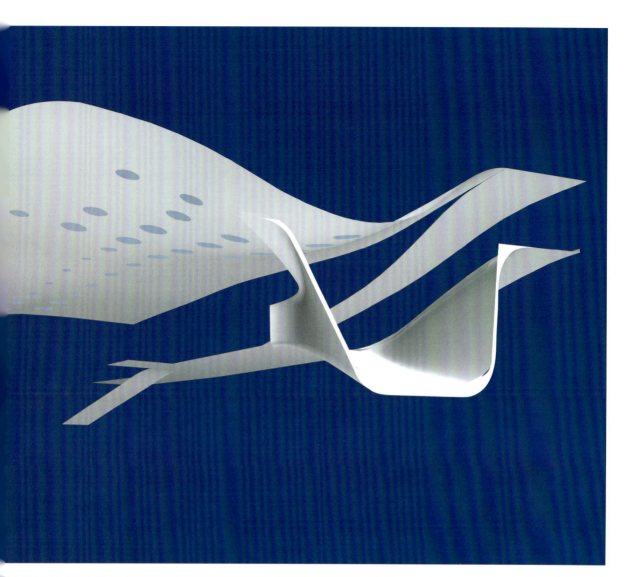

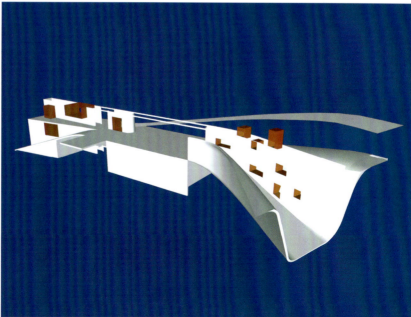

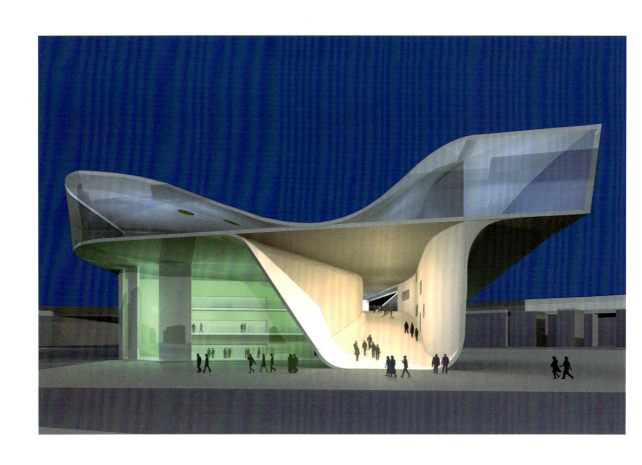

上图：形同乐器的建筑
above: a house like an instrument

右图：交响乐之旅的路径
right: routes on a symphonic journey

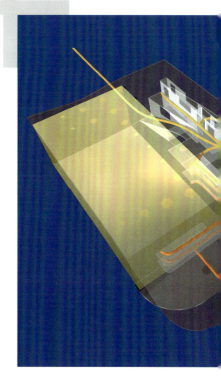

特别为当代音乐设计的建筑便成为本次设计的中心理念。计划中的音乐厅将由多种技术支撑,并且独到的声学设计,可以使演员和听众按照最佳声音效果或听音经验自由地选择位置。

在音效和结构工程师的配合下,一个"房中之房"的概念逐步产生。这个被叫做"黑匣子"的房间是一个灵活的空间,可以根据不同的表演需求在三维方向伸展和收缩。

另外,还有研究实验室、展示空间、媒体图书馆和创作排练舞台。这些在市中心吸引人的地方都对公众开放。为了实现这一观点,设计者按照行人的步行路线在外部设置了玻璃幕墙,行人能够顺着它穿过并到达建筑各楼层,到达屋顶并又返回,让人们一开始便能感受到当代音乐的魅力。

A specially designed hall for contemporary music was at the centre of this concept. It was intended as a hall of technical variability and acoustic uniqueness in which cast and audience alike could freely choose their place for an optimum projection of sound and listening experience.

In cooperation with acoustics and structural engineers a concept of a "room within a room" was developed. This so called "Black Box" is flexible and can be extended or reduced three-dimensionally meeting the various requirements of different performance practices.

In addition research laboratories, exhibition spaces, a media library and a workshop rehearsal stage were meant as an attraction at the centre of town, open to the public. Further essential elements in support of this idea were the outer glass shell as well as a pedestrian foot link through and across all levels of the building, up to the roof and back down again, allowing for people's initial contact with contemporary music.

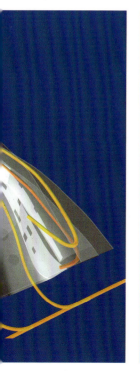
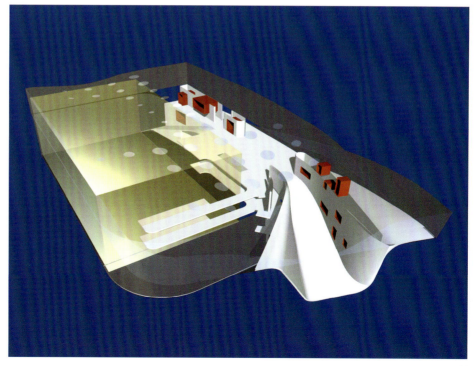

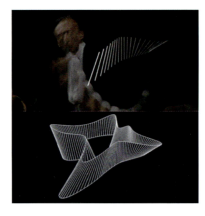

日本2005年世界博览会德国展馆，日本爱知县，2004年
一次穿越德国的象征性旅行

设计：格瓦斯·昆·昆
合作设计：ART+COM and Pearlman

在日本2005年博览会为德国馆而作的这个设计，主要是想富有诗意地表达一种代表德国的创新价值观。通过与ART+COM和Pearlman的合作，产生了一个设计德法孪生展馆的独特的展示概念，它包括建筑室内和入口形象。

设计方案主要有两部分：一个被声学幕墙包裹的大厅和一个和谐流动的室内空间，都是根据音乐波动与振动的规律进行设计的。镶有饰片的透明立面由无数木制圆片组成，微风吹过能使他们发出响声。

GERMAN PAVILION AT THE EXPO 2005, AICHI/JAPAN, 2004
A SYMBOLIC JOURNEY THROUGH GERMANY

Design: Gewers Kühn + Kühn
in cooperation with: ART + COM and dan pearlman

The design for Germany's national presentation at the EXPO 2005 in Aichi, Japan, was for a poetic translation of the country's innovative values. In cooperation with ART + COM and dan pearlman an individual exhibition concept for a decreed German-French twin pavilion was developed, which embraced both the interior and the entrance façade.

There are two main components, an acoustic curtain façade enveloping the hall and a harmoniously moving interior, which both follow the gentle laws of musical waves and vibrations. The transparent sequin façade consists of countless round wooden pixels which start to

上图：指挥棒的运动轨迹
above: movement of a baton

右图：转化成奇特的空间概念
right: translation into a special concept

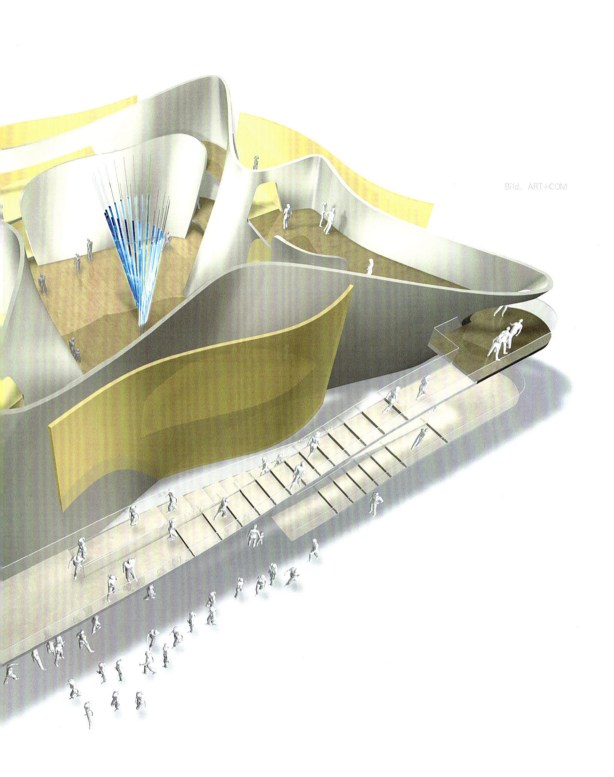

Bild: ART+COM

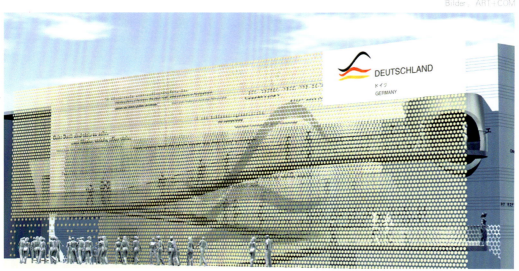

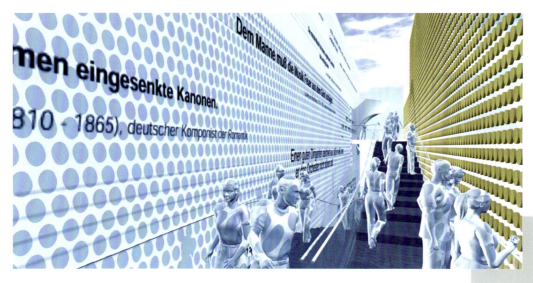

右上图：楼层平面
above right: floor plan

左上图：能发声的入口外观
above left: sound generating entrance facade

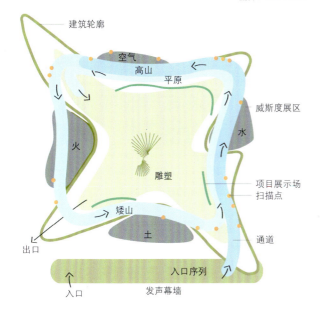

微妙而充满诱惑的特点将吸引参观者进入室内一探究竟。

人们沿着一条通往展馆主体和声学幕墙之间的步行通道进入展馆，当人们沿着通道向前行走时能听到由德国音乐家演奏的音乐旋律和诗意片断。室内的形状是通过分析指挥家指挥棒的舞动路径得来的——多样的片断，外露的激情。指挥家那流动的创造性而又自制的表达方式成为展馆品质的有趣范本。这里的展览空间是一种不可再分的交响乐空间，一种宁静的、互动的空间。舞动的形态避免了物质层面的机械设计而达到了精神层面的水平，使人们更加贴近地感受德国。

sound when gently moved by the wind. This subtle but enticing feature invites visitors to explore the interior.

Access to the pavilion is gained via a footpath rising up between the curtain and the pavilion box. On their way up people walk through a cloud of sound of melodies and fragments of poetry performed by German musicians. The interior is shaped in analogy to a conductor's moving baton – in multiple segments, impulsive. His dynamic, creative, yet controlled way of working serves as a congenial example of the qualities required for this pavilion. The exhibition area is an indivisible symphonic space, a serene, interactive spatial choreography devoid of objects, which appeals on an emotional level and seeks to make people feel closer to Germany.

奥林匹克建筑综合体"体育的动感",莱比锡,2003年
活动的建筑

"体育的动感"是活动的建筑的最初理念。为了支持莱比锡申办2012年奥林匹克运动会,这座建筑将作为2006年世界杯足球赛的新闻与展示中心,同时它还将在奥林匹克运动会期间提供信息。建造一个可以移动而不是静止的建筑将是一个巨大的挑战。这座建筑本身表现了体育作为事件的观念,它将比常规传统的展览建筑更具吸引力。

透视图
perspective view

OLYMPIC BUILDING COMPOUND "SPORT MOVES", LEIPZIG, 2003
MOVING ARCHITECTURE

"Sport Moves" was the initial idea for an architecture in motion. In support of the City of Leipzig's bid to host the Olympic Games 2012 the building was to accommodate the press- and exhibition centre for the 2006 FIFA World Cup and at the same time offer information on the Olympic Games. The challenge was to create an architecture that physically moves instead of being static. This building itself was to express sport as an event and an attraction beyond all conventions of traditional exhibition halls.

这个"体育的动感"设计概念包括两个元素："桌子"和"指环"。"指环"是流线型四分之三周长的镰刀状构筑物，它可以靠液压动力在对角线方向上下移动，为观赏奥林匹克场地提供最大的景观视角。通过运动在这块体育场地上展示体育的理念。"指环"内部为奥林匹克委员会提供独一无二的工作空间，并且为参观者提供了观看各种表演和展览的场所。

利用航天制造技术和造船技术中的现代空间结构技术，"指环"内部的无柱空间成为可能，而且作为结构整体只需计算受力中心位置的自重。另一方面，"桌子"部分作为入口的特殊展示空间，安排了所有承载较大的功能空间，如永久性展馆、仓库、储藏室。它与"指环"有着同等重要的地位。

纵观整体，整座建筑给人一种有关体育和所有运动的独一无二的震撼的感受。

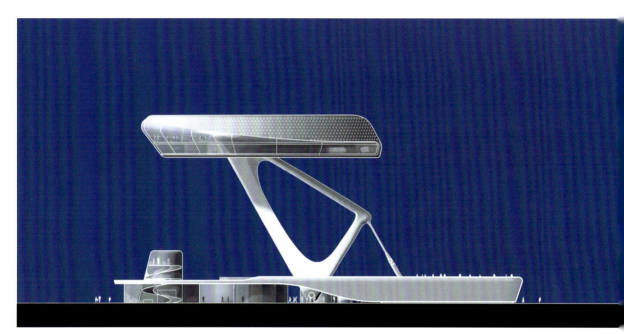

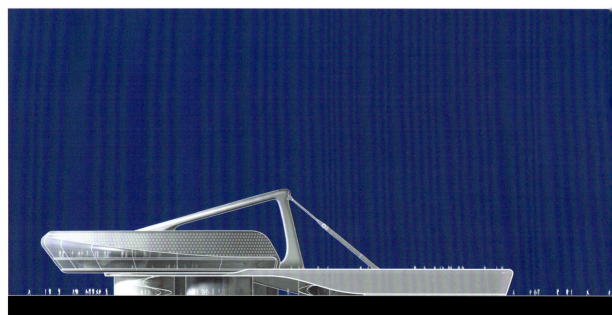

The design concept of "Sport Moves" was based on two elements, a "Table" and a "Ring". The "Ring" was a dynamic three-quarter length sickle shaped structure which could be moved up and down diagonally by hydraulic presses offering maximum panoramic views of the Olympic grounds, representing the idea of sport as movement on the sports ground. The "Ring" inside was designed to offer a unique work place for the Olympic community and a venue of alternating shows and presentations for the visitor.

Using modern Space-Frame-Technology as applied in aircraft construction and ship building the "Ring" offered the possibility of a column-free interior taking into calculation the structure's own weight at the centre of gravity. The "Table" on the other hand accommodated all heavy-weight installations such as the permanent exhibition, the depot, the stock room or special exhibitions as an attractive entrance area and a counterpart to the "Ring". Together the ensemble would have offered a unique and immediate experience of sport and all facets of movement.

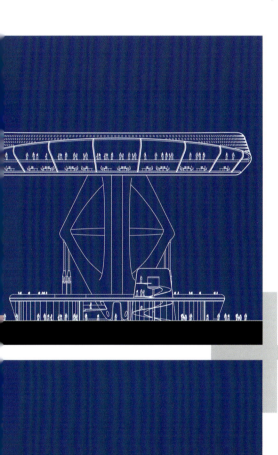

左图："指环"的上下运动
left: the up and down movement of the "Ring"

右图：总平面
right: site plan

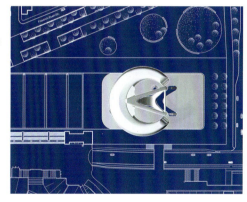

医疗科技博物馆，埃尔兰根，2003 年
看起来像细胞结构的建筑

过去与未来的有机共生是埃尔兰根医疗科技博物馆的主要教育理念。在这个曾经生产过第一台X光机的工厂厂址上，埃尔兰根市决定对产业结构进行调整，以使这座城市由原来著名的电子产业基地转向媒体和医疗健康产业。为了这个目的，老厂房将被改造成适应未来多学科发展使用的结构，大量仿生学设计的形象将被应用到新老建筑中。

一种自然现象成为这次改造的灵感来源，即人类长骨中非常有用而且具有美感的细胞结构，这种结构形式应用到建筑中，不仅能够用很少的材料产生强大的支撑结构，而且会创造出新颖奇妙的空间形式，它将挑战人们的想像力并唤起人们的好奇心。

MUSEUM OF MEDICAL TECHNOLOGY, ERLANGEN, 2003
LOOKING AT CELL STRUCTURES

A symbiosis of the past and the future was the educational concept for the Museum of Medical Technology in Erlangen. At the exact site of the first factories for X-ray machines the city of Erlangen intended to launch a structural change from a city renowned for electronical engineering to a place of innovation in the fields of the media and health. For this purpose the old factory is transformed into a structure enabling many interdisciplinary interactions. Numerous biomorphological interconnections merge the old and the new.

A natural phenomenon is the point of departure for this transformation: the extremely efficient and aesthetic cell structure of human long bones. Their macroscopic adaptation into architecture enables the construction of unusually

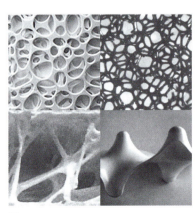

上图：博物馆入口
above: entrance of the museum

左图：作为结构灵感的
分子结构图
left: molecular structures as inspiration for the construction

上图：全景表现图
above: perspective view

右图：无缝建筑
right: seamless architecture

如果我们把现存建筑内部的固定装置和其他结构全部掏空，那么这座建筑将成为一个支撑混凝土外壳和拱梁的巨型生物器官结构。已经封闭的工厂也因此而对城市开放。

通过显微镜和内窥镜看到的图像被打印到玻璃外壳上，生动而透明。建筑装备有低能耗的表皮和丝制的印花帷幕，表皮成为一个能量过滤器，同时日光屏幕作为媒体信息演示版使用。室内巨大的斜坡迂回穿梭于相互交错的有机体中，把它们组织成连续的空间。

strong supporting structures with only a minimum use of materials and the creation of new, fascinating spaces, which challenge our imagination and awaken our curiosity.

After gutting the existing building from all fixtures and extensions it was developed into an organic macro-structure of load-bearing concrete shells and arched girders. The formerly hermetically sealed factory site thus opens up towards the city.

It has an inviting, transparent look about it with images from the world of microscopy and endoscopy printed onto its glass shell. Equipped with a low-energy coating and silk screen printing (serigraphy), the façade acts as a gentle energy filter and daylight screen and at the same time performs as a media information board. The interior is organised as a single space continuum in which large ramps literally weave their way through an interactive organism.

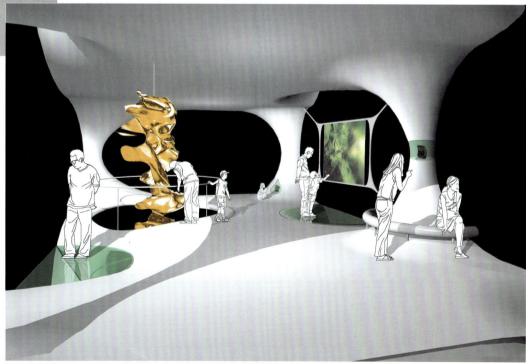

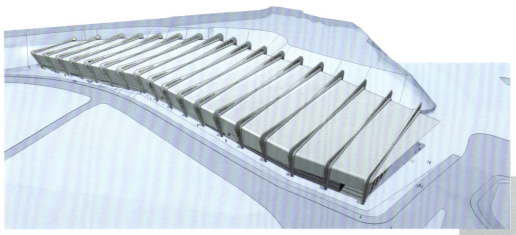

上图：总平面
site plan

下图：朝向城市的立面
below: elevation towards the city

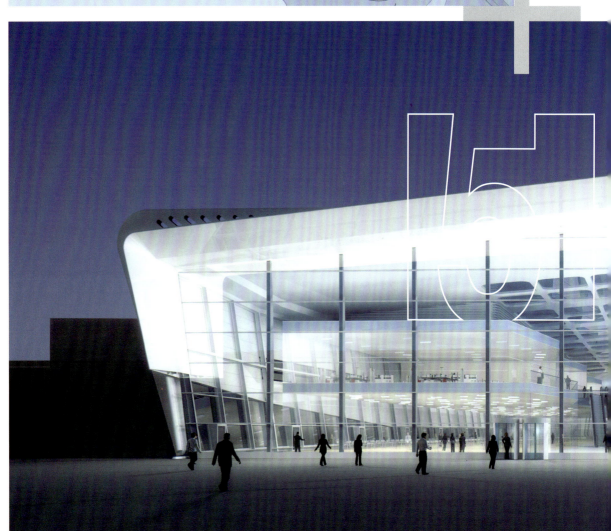

THYSSENKRUPP 学院，埃森，2006年
钢的反思

THYSSENKRUPP ACADEMY, ESSEN, 2006
STEEL – RETHOUGHT

ThyssenKrupp学院这个典雅的金属壳体被设计作为一个多功能的文化场所，包含一个音乐厅以及研究和技术设施。为了颂扬钢铁公司对这座建筑和波鸿这个钢铁城市名声的影响，钢是这个设计创意的优选建筑材料。

This elegant metallic shell for the ThyssenKrupp Academy was designed as a multiple-use cultural venue housing a concert hall as well as research and technology facilities. Paying tribute to the steel company lending its name to this edifice and to Bochum – the city of steel – steel was the preferred building

上图：单元翼
above: wing unit

我们与著名的工程师 Werner Sobek 合作设计出一个轻质大厅，这个大厅自由而灵活地体现了"房中之房"的概念。建筑的一部分斜向一边，另一部分抬起，因此形成了新公司总部中心引人注目的标志。这个建筑的主要构架是一个倾斜的三角铰接梁，跨度70m长，轴间距15m。结构使用最少的材料并且完美地遵循力的流动。

现存的非匀质的景观被纳入到大厅之中，巨大的建筑体量非常完美并且通过一个和建筑物一样高的引人注目的大厅向城市打开。在内部，所有的生活区域被自由地安排在巨大明亮的钢结构下面。建筑体量和空间被融入进一个新的城市独立空间中。ThyssenKrupp 学院大厅提供多用途使用。这里，位于公司中心的这个新建筑标志着我们的社会从工业时代到知识时代的前进。

左图：桁架结构的钢构件
steel components of truss structure

摄影：Ralph Baiker

material for this innovative development. Together with the renowned engineer Werner Sobek a big weightless hall was developed which freely and flexibly incorporates the concept of a "house-within-a-house". Part of the construction slopes downwards on one side and rises up on the other, thus forming an inviting technical landmark at the centre of the new company headquarters. Its primary construction, an inclined triple hinged girder, spans 70 metres in length, the distance between the axes measuring 15 metres. It is constructed with a minimum use of materials and elegantly follows the flow of forces.

An existing unevenness in the landscape has been integrated into the hall, the large volume has been rounded off and opens up towards the city with an inviting glass foyer as tall as the building itself. Inside, all utility areas are arranged freely under the large shining steel construction. Objects and spaces are merged into a new independant urban space. The ThyssenKrupp Academy Hall is intended for multiple use. Here, situated at the very heart of the company, this new construction marks the evolution of our society from the industrial age to the age of knowledge.

右图：内部空间
interior space

漂浮之家，马尔马拉海岸，2003 年
小行星馆

FLOATING HOME,
MARMARA COAST, 2003
BUILDINGS FOR A SMALL PLANET

考虑到日益增长的世界人口，不足地球表面 30% 的陆地资源显得愈发宝贵。我们的子孙后代将来能在哪里居住？拦海拓地的现代方法消耗了大量的能源，与各种限制因素背道而驰。

As little as 30 percent of the earth's surface is land, an increasingly precious resource in the light of the rapid increase of the world´s population. Where will our future generations find space for living? The modern way of reclaiming land from the sea consumes huge amounts of energy and is beginning to come up against

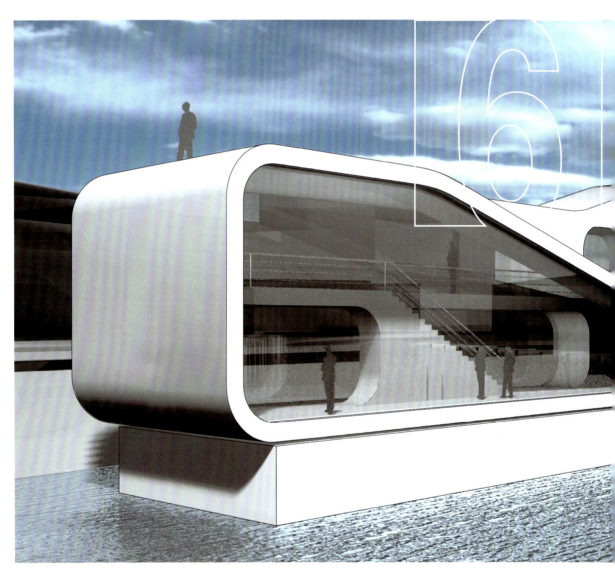

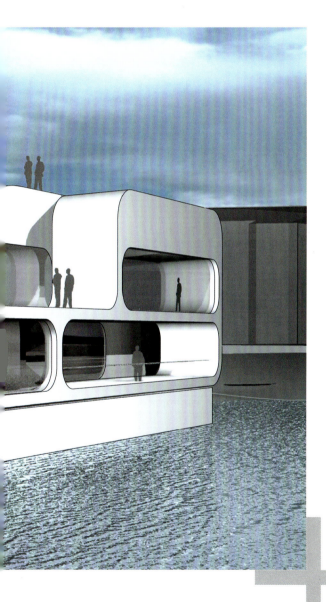

漂浮之家
floating home

41

数千年来，人们已经在水边或水上建造住宅。难道现在不正是采用新方式、并开发新的建造方法，在水上建造房子的恰当时机吗？

连同一个大的土耳其船坞，"漂浮之家"被设计出来了，一个以模块为单元的水上住宅，像海上的船队一样。它可以用作居住、工作和娱乐，并且符合可移动和灵活的需求。作为货物集装箱的后勤基地，适用于船舶、飞机、火车和货车，"漂浮之家"的钢制模块单元可以以多种方式进行整合，然后重新安置在水上，创造各种新的活动空间。

"漂浮之家"不只是普通的"船屋"——它能在世界各地任何一条河畔或海岸为人们提供现代的居住、娱乐和办公设施。根据最初的市场研究，在斯堪的纳维亚、马尔马拉海（土耳其西北——译者注）、佛罗里达以及克里木（乌克兰东南部黑海中的一个半岛——译者注），这种房屋的需求量十分巨大。由于它那充分利用水和风能资源的完美理念，"漂浮之家"将会是一个在更大尺度上通过结合运用取之不尽的水资源和风能，回应21世纪的巨大挑战。

limiting factors. For thousands of years people have already built houses in areas by and above the water. Isn't it high time to tread new paths and start developing new construction methods for buildings on the water?

Together with a large Turkish shipyard the "Floating Home" has been developed, a modular system for houses on the water, which – like sea caravans – may be used for living, working and recreation, as required, in a totally mobile and flexible way. Based on the logistics of freight containers, applicable for ships, planes, trains and trucks, steel modules have been developed, which can be assembled in a variety of ways and then be relocated and rearranged in order to create new spaces for activities on the water.

The "Floating Home" is much more than just an ordinary "houseboat" – it provides people with modern living, recreation and office facilities on river banks and sea shores all over the world. According to initial market research such homes are currently in great demand in Scandinavia, on the Marmara Sea, in Florida and on the Crimea. With its integrated concept of using the sheer inexhaustible resources of water and wind energy, however, the "Floating Home" could be an answer to the great challenges of the 21st century on a much larger scale.

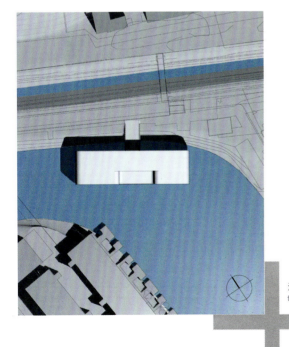

漂浮的办公室
floating office

无尽的高楼
endless tower
右图:航拍合成图
right: aerial view

地标建筑与水下酒店,青岛,2007年

Hydropolis是一种先进的建筑概念来开发豪华水下宾馆,这正如我们感受到的在世界范围内的一些特别的滨水城市如迪拜、摩纳哥、悉尼和中国的青岛。

第一个项目将会在中国实现。青岛引人入胜的海岸线已被选定为一个重要的地点,2008年奥运会水上赛事将在这里举行,并已开始了一些重要进展。

LANDMARKBUILDING WITH UNDERWATER HOTEL, QINGDAO 2007

Hydropolis is an advanced building concept developing luxury underwater hotels which will be realised worldwide in specifically chosen waterside cities such as Dubai, Monaco Sydney and Qingdao/China.

The first project will be realized in China. The spectacular coastline of Qingdao has been choosen as an important location because the sailing race of the 2008 Olympic games will take place here and is already initiating several important developments. The land station as seen on the perspectives, is the first building phase incorporating the reception, the entrance and lounge for the underwater hotel, own restaurants, a shopping arcade as well as luxury housing at the waterfront and in the tower.

正如效果图中所看到的地面站，是建筑的第一期位于海滨和塔中，其综合了接待处、入口、水下酒店休息厅、特色餐馆，购物商店以及豪华住宅。

建筑本身也可以从陆地部分进入，也可以从水上进入。一旦进入建筑中，一个豪华的穿梭运输工具会将酒店的顾客带入到水下。

建筑物表现了它潜在的动态性和外在形态，其连续的水平和垂直的运动表现在其具有部分多孔的立面有机表皮。

The building itself can be entered from the landside as well as from the water. Once within the building a luxury shuttle will take the hotel guests under water.

The buildings architecture shows its dynamic potential and portraits its continuous horizontal and vertical movement and reminds with the partly perforated facade of an organic skin.

Gewers Kuehn + Kuehn Architects Berlin 2007

建筑理念
architectural concept
右下图：最精彩的部分
below right:
highlights

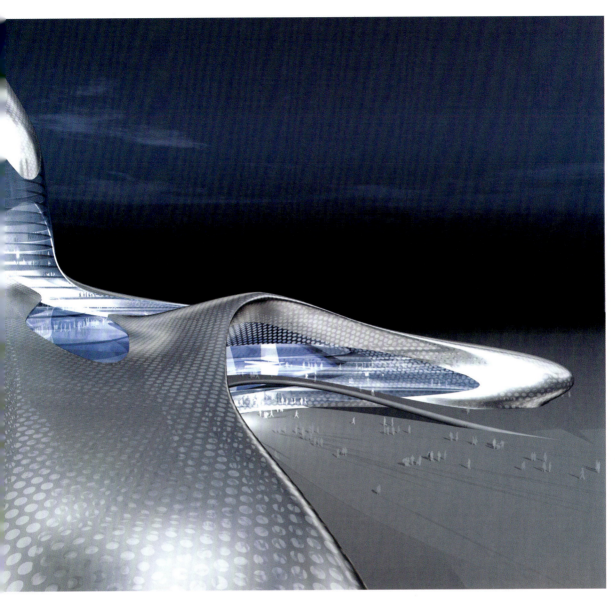
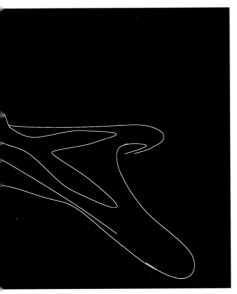
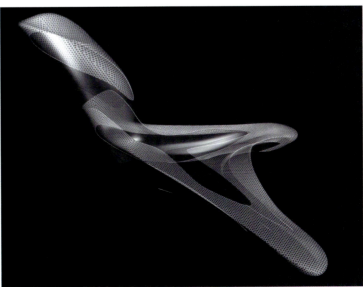

帕莱斯垂直"花样住宅",波兰,华沙,2004年
"城市之春"

PALAIS VERTICAL "BLOSSOM HOUSE", WARSAW/POLAND, 2004
„SPRINGTIME OF A CITY"

作为东欧复兴和华沙都市活力的标志,这座高层综合楼主要是作为公寓来开发的。建筑用地毗邻华沙中心火车站,一座展示斯大林式王冠的建筑。设计意在表达新的年轻一代的生活方式。在华沙城保存较好的建筑中,既有色彩丰富的装饰性建筑,又有战后整齐划一的装配活动房。"花样住宅"试图传达一个积极的信息,那就是一切差别都可以克服。

As a sign of the emergence of Eastern Europe and the liveliness of the metropolis Warsaw this high-rise hybrid was developed to accommodate mainly apartments. Here, in direct vicinity of Warsaw's central railway station displaying the Stalinist city crown, the design incentive was to express the lifestyle of a new and young generation. Amidst the fine urban structures of old Warsaw with its colourful ornate buildings on the one hand and its uniform post-war prefab structures on the other, "Blossom House" was intended to put across a positive message, that all differences can be overcome.

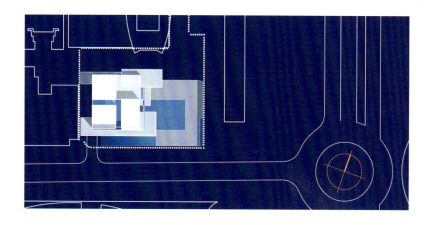

左下图:华沙市中心总平面
below left: site plan city centre Warsaw

花样住宅
blossom house

第一步，150m高的建筑主体被雕刻成流线型，空洞从体量上挖出使塔楼深处尽可能多的得到光照。每一层可以容纳最少八套灵活布置的公寓房。

第二步，那些空洞的出现被想像成连成一体、形成一个特殊的垂直通道，部分公共、部分私密的不同房间垂直组织形成的景观，将周围城市生活与这座高层建筑紧密地连在一起。艺术画廊、壁球馆以及悬挑的屋顶花园等等体现了这座高层建筑开放空间的多样性。

最后一步，明亮而又色彩丰富的特殊多层玻璃体块设计成夹在塔楼中冒出的样子，犹如一簇从枝干上盛开的鲜花。它们共享同一个主塔楼，但在颜色和形态上却各有特色。它们在视觉上表达了华沙的新精神。新的城市活动以及个性感的出现将带来城市的复兴。

In a first step the 150 metre high base structure was to be sculpted into a dynamic shape and "voids" cut out of the volume in order to guide as much light as possible deep into the tower. Each floor allows for the construction of at least eight apartments of flexible sizes.

In a second step these emerging voids were supposed to be linked to form a unique vertical passage, a vertically organised landscape of different rooms, partly public, partly private, closely linking the high-rise structure with life in the surrounding city. Art galleries, squash courts, clubs or suspended roof gardens were intended as variants opening up the high-rise structure.

左下图：多彩的公寓楼如花丛一样
below left: colourfull multi-storey appartments like blossoms

In a last step special bright and colourful multi-storey glass cubes were supposed to emerge from the tower like clip-ons, similar to the way the blossoms of a flower shoot from the stem. Individual in colour and shape, yet all sharing the same base structure, they would have visibly expressed the new spirit of Warsaw, a rejuvenated city with new urban activities and an emerging sense of individualism.

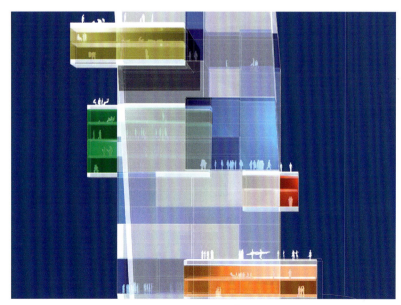

剖面图
section

完全购物中心,埃森,2005年
文脉的重生

  在埃森繁忙的市中心的中部,有许多条快速车道,一条到达新购物中心的简捷道路以醒目的方式将这个购物中心和城市整合为一个整体。其结果是一个完美的设计项目,它不同于所有的传统的零售设计方式。它以奇思妙想吸引了沿线的司机和行人。

  购物中心设计成流线型,在一个多边形立面的后面,购物和停车设施合并成一个整体。由青铜色经阳极电解的铝制作的三角形板的网状面壳沿着对角线方向铺设,使建筑产生一种三维的立体感。它与周围一般的环境如此不同,令人耳目一新。

52

鸟巢结构的形象
a netstructure
as façade

TOTAL SHOPPING, ESSEN, 2005
REGENERATION OF THE CONTEXT

In the midst of Essen's busy city centre with its many fast traffic lanes the brief was for a new shopping centre to be integrated in an attractive and visually striking way. The result was a precious design object denying all conventional retail design features, a curious attraction to drivers and pedestrians alike.

It was developed as a dynamic visual body. Behind a polygonal façade shopping and parking facilities merge into one unit. A web-like shell of triangular panels made of bronze-coloured anodised aluminium is laid at a diagonal angle, thus lending a three-dimensional presence to the building. It is refreshingly different and stands out from its rather ordinary environment.

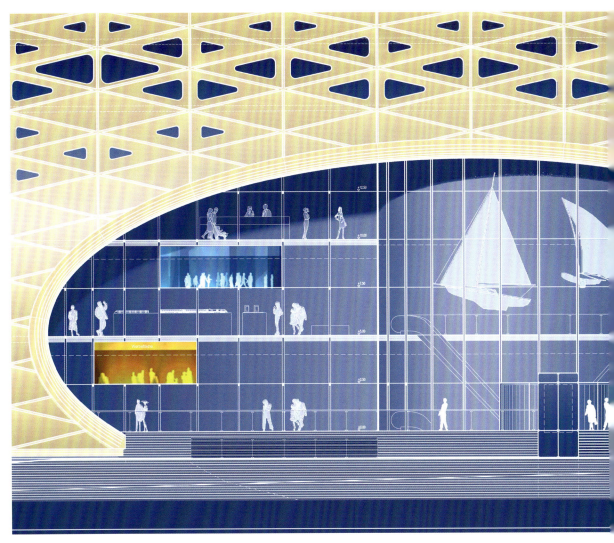

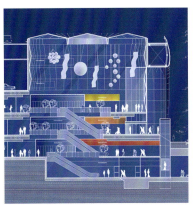

这个巨大的建筑体块只需通过改变外部面板的开启角度就可以轻而易举地改变其外观效果。从不透明到半透明再到完全透明之间的许多效果均可实现。在仅有的几个全玻璃立面中,广告被减少了,设在购物中心大橱窗后面,并被有效整合到建筑中去。恰好高于视线的间接照明创造出一个大型建筑盘桓在城市上空的戏剧性效果。

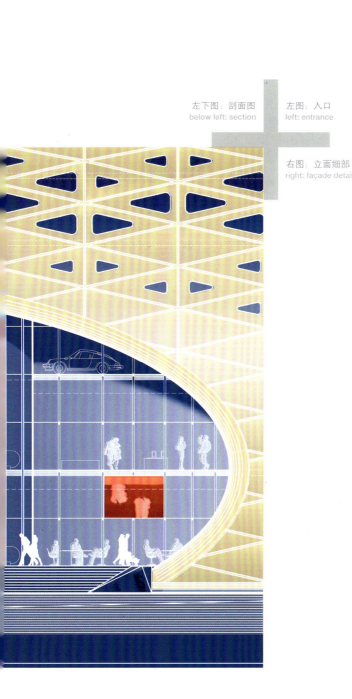
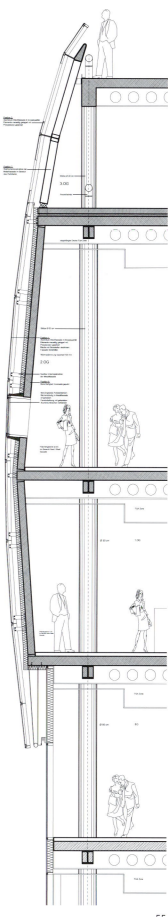

左下图：剖面图
below left: section

左图：入口
left: entrance

右图：立面细部
right: façade detail

The large building mass can change its visual appearance simply by varying the panels' opening angles. A range of looks from opaque to translucent or transparent is possible. Advertising is reduced to the few all-glass sections. Displayed behind large shop windows it is effectively integrated into the architecture. Indirect illumination just above eye level creates the dramatic illusion of a large building unit hovering above the city.

奥迪公司变速箱及排放研究中心，英戈尔施塔特，2005年
"机器里没有多余部件"

AUDI LTD GEARBOX AND EMISSIONS CENTRE, INGOLSTADT, 2005
"A MACHINE HAS NO SUPERFLUOUS PARTS"

为奥迪公司设计位于英戈尔施塔特新的变速箱及排放研究中心大楼时，一项目标以建筑表现汽车的创新技术：在一个集中的建筑中安排复杂的功能用房。诸如测试台、实验室和杂物间之类的区域必须与管理区域联系紧密，而不同区域间互不干扰。奥迪公司铝合金制变速箱的优雅与高效的形象用在建筑设计中同样适合。

结果形成了一座刻意表现高效企业文化的复合建筑，"白领"区与"蓝领"区在这里得到平衡与协调。这座建筑图式清晰且简洁，所有的技术部门被安排在底层。

Expressing innovative automobile technology in terms of architecture was the objective when designing the new Audi Ltd Gearbox and Emissions Centre in Ingolstadt: a complex room scheme concentrated in a compact building. Different areas such as testing stands, laboratories and garages had to be closely linked with administration areas and offices without interfering with one another. The elegance and efficiency of Audi's aluminium gearbox was to find its equivalent in architecture.

The result was a hybrid building representating an efficient and design-conscious corporate culture where "Blue Collar" and "White Collar" areas are coordinated as equals and in harmony. The building diagram is clear and simple. All the technical departments are

下图：总平面　　透视图
below: site plan　　perspective view

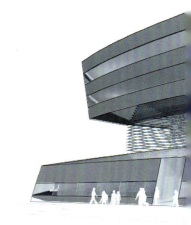

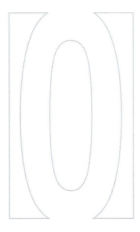

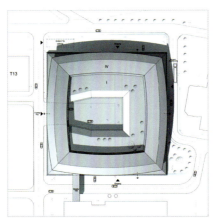

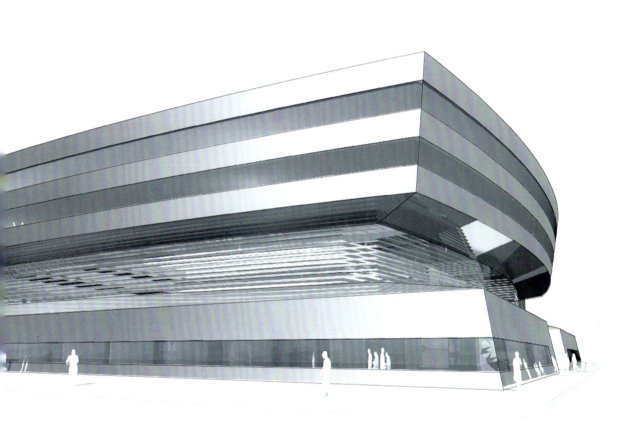

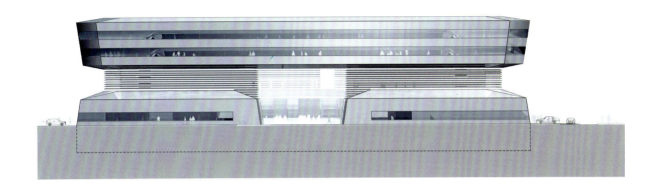

西南立面图　剖面图
elevation south-west　section

然后有两层呈环状的办公层，流线型设计挑出了建筑结构轴线。在这两个区域之间，巨型建筑结构技术创造出一个大型室外平台。平台为公司员工提供了一个交流的空间，改善公司内部的协作关系。

为不同元素进行的设计以当代汽车技术作为原型：建筑主体是由银白色金属板覆盖着的底层和环状办公室构成，精制的水平铝合金百叶板窗好像在这栋建筑里开发和测试的圆柱形机器。视觉冲击力和高效率使这座建筑不仅仅是一个工业建筑，它完全是一个表达奥迪创新企业文化的优秀作品。

situated in the base unit. Then there is a two-level ring construction with offices, its dynamic shape projecting beyond the actual building structure. In between these two areas, the massive building technology is accommodated with a large outdoor terrace as a special place for communication intended to improve synergies within the company.

The design for these different elements is modelled on contemporary automobile technology: the bodywork is made of large silvery-white metal panels covering the base unit and the ring, precise horizontal aluminium louvers resemble the cylinder of an engine just like those developed and tested in this very building. Visually striking and efficient, this is not merely an industrial building; it is a precious workpiece representing the innovative corporate culture of Audi.

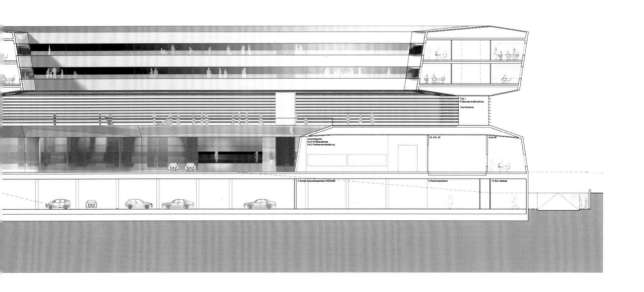

中央火车站新站，慕尼黑，2006年
21世纪的火车站

位于慕尼黑市区的中央火车站新站意在创造一个作为城市景观组成部分的多功能和多维度的建筑。设计考虑的主要标准是都市整合

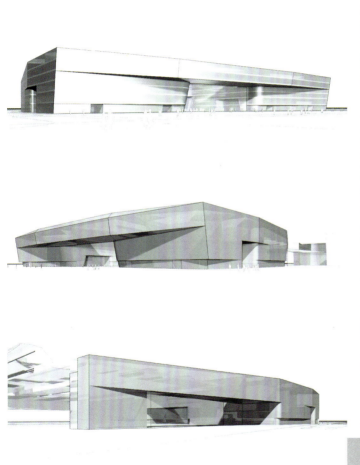

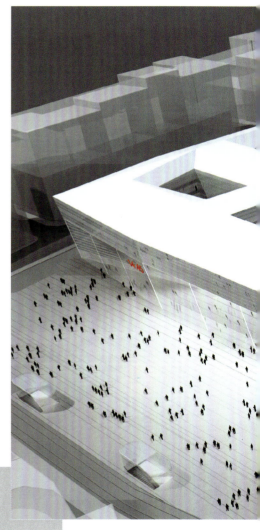

城市文脉中的雕塑
sculpture in an urban context

化，可持续发展程度和效率。根据实际情况，我们既需要小心对待现存的建筑结构，又需要对它们进行毫无保留的重新计划，以适应公众对 21 世纪的车站的需要。

与开发基础设施的概念不同，设计聚合了现有的不同时代的形式，并将其转换成一种美学的和功能化的单元。

NEW CENTRAL RAILWAY STATION, MUNICH, 2006
A RAILWAY STATION FOR THE 21ST CENTURY

The design brief for this new central railway station in the city of Munich was to create a multi-functional and multi-dimensional building as a constituent of the city landscape. Urban integration, sustainability and efficiency were the main criteria to be considered. In practical terms this meant to treat the existing structures with care and to reprogramme them unreservedly to meet the public's demand for a station fit for the 21st century.

A differentiated concept for the development of the infrastructure extends to the existing conglomeration of styles from different eras and converts it into an aesthetic and functional unit. A single transparent volume bathed in daylight

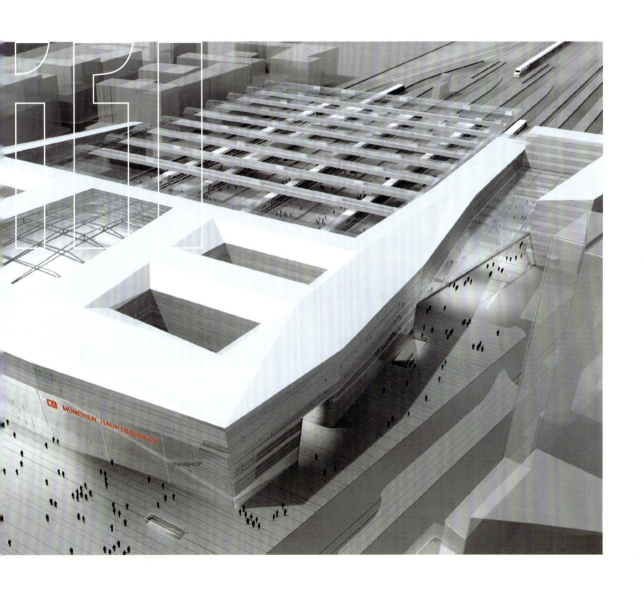

61

沐浴在日光中的单纯透明体提供了 35000m² 的办公空间，25000m² 的购物设施和 18000m² 的旅馆。

这些开发计划分阶段实施，占据着新的城市空间，并将火车站与城市空间整合为一体。从火车站深处延伸出一个平台，伸展到新火车站前院，继而直接延续到附近的 Stachus 城市广场。在入口大厅内设有螺旋链式的构筑物，以各种方式将位于街道 40m 深度的各层平台联系起来。这个车站广场明亮宜人，易于识别路线。漫步在以流动和连续的方式连接的空间序列中，是愉快和惬意的。

当代轻质、透明的建筑材料将这座车站变成一个真正的现代公共交通建筑，流动的造型为车站内部和它所在的城市增添了活力。

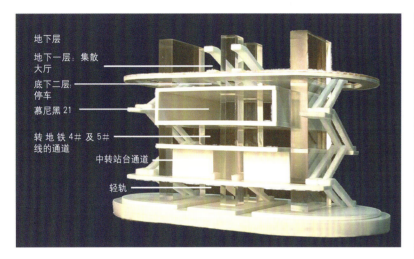

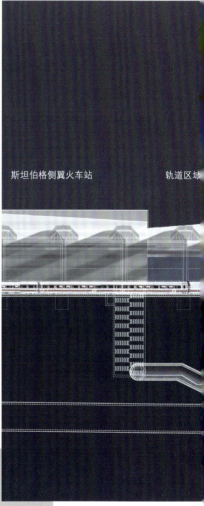

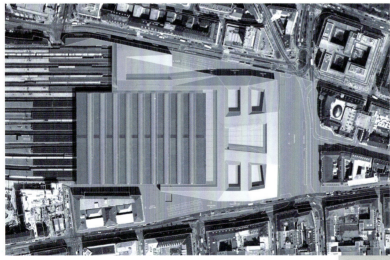

左上图：中心细胞
above: nucleus

右图：剖面图
section

左下图：城市的流动
below: urban flow

accommodates 35.000 square metres of office space, 25.000 square metres of shopping facilities and 18.000 square metres designated for a hotel. These developments, to be realised in phases, are to occupy new urban spaces and to re-integrate the station into the city.

Emerging from deep inside the station itself is a plateau spreading out into the new station forecourt and then directly continuing on to the Stachus, a close-by city square. Inside the entrance hall there is a spiral-like construction in the manner of a nucleus which creates various links to all platform levels situated as deep as 40 metres below street level. It is easy to find one's way round this pleasant and bright station concourse. One's stay is made even more agreeable and lively by wandering through the sequence of spaces which are linked in a dynamic, continuous way. Contemporary light-weight, transparent building materials turn this station into a truly modern public transport building. Its dynamic body pays tribute to both the movements inside the station and to its location in the city.

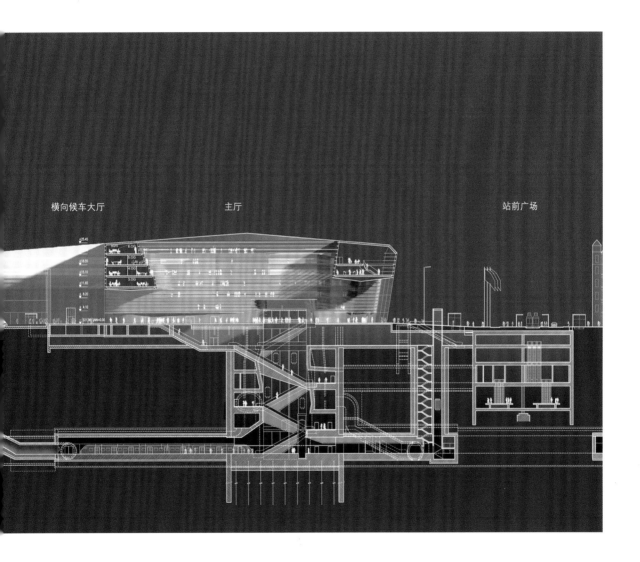

左图：主大厅景观
view into the main hall

右图：面向城市中心的主立面
main elevation facing the city centre

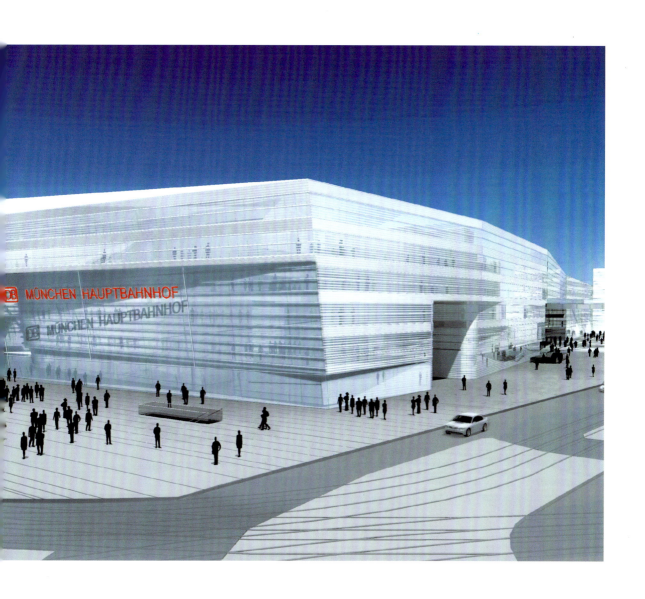

白南准博物馆，韩国京畿道，2003年
河床上的卵石

一个新的博物馆概念为韩国著名的概念艺术家 Nam June Paik 创造了展示动态视听艺术作品的独特陈列空间。他的构想是非等级的、清晰的、动态开放和容易接近的，公众易于接受将外部与内部紧密连接的"移动的图景"。

考虑到馆中陈列的艺术家的作品和环绕博物馆的区域，即以森林为主的山谷，博物馆构想被具体化：建筑呈流动的体形，并提供不同的参观路线。它独特的空间流动性和均匀的形状像一粒在水中的卵石。博物馆的建造技术都是创新的，就像白南准的作品一样。

NAM JUNE PAIK MUSEUM, KYONGGI/ SOUTH KOREA, 2003
LIKE A PEBBLE IN A RIVER BED

A new museum concept has been developed for the famous Korean concept artist Nam June Paik creating unique exhibition spaces for his dynamic audio-visual works of art. His conception is non-hierarchical, clear, dynamically open and accessible, allowing the public to be swept away by his "moving images" which closely link the interior with the exterior.

In consideration of both the artist's work and the area surrounding the museum, a valley framed by forests, the idea of a museum as a moving body resting in itself and offering different routes crystalised. Its unique dynamic and well-balanced shape resembles a pebble in the water. Its construction is a result of truly innovative technology just like Nam June Paik's work itself. The outer shell made

无缝建筑
seamless architecture

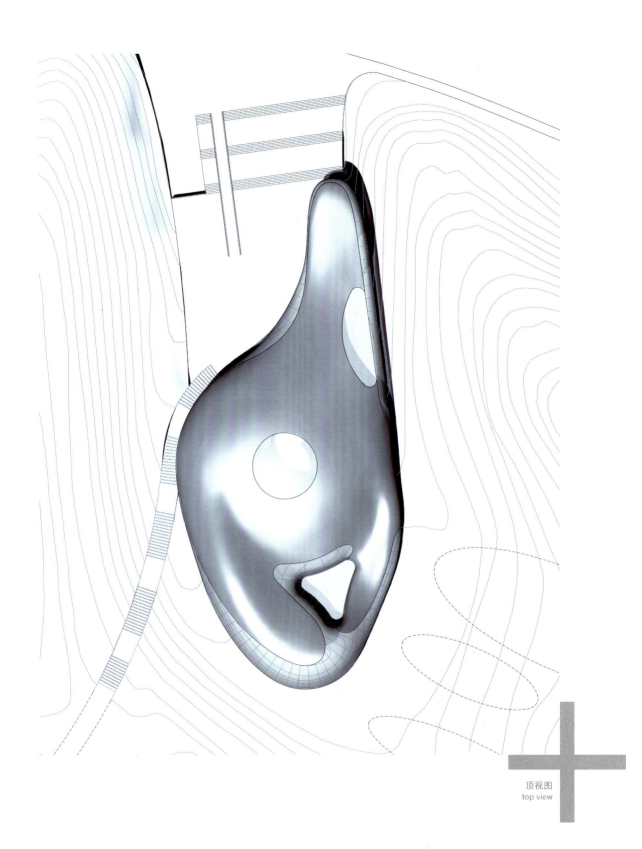

顶视图
top view

建筑外壳由纤细的玻璃纤维制成,部分透明,部分半透明或不透明,适于向远处展示大规模的视频展览。

在中央大厅周围自由流畅地环绕着12个相等且灵活的展览区域,就像卫星盘旋在轨道上。参观者可以自由地接近坡道上展出的白先生的作品。他的虚幻的艺术世界在与周围山谷的真实世界进行着持续不断的对话。博物馆表现出没有物质界限的空间和连续介质的自发性和流动性,编织着理性和情感相互微妙结合的网络。

of thin fibre glass, partly transparent, partly translucent or opaque, has the facility to show large-scale video displays visible from afar.

Inside there are twelve equal and flexible exhibition sections freely and fluently grouped around a large central lobby like satellites circling it in orbit. Visitors may freely access Mr. Paik's works which are exhibited on ramp-like levels. His virtual artistic world can stand in constant dialogue with the real world of the surrounding valley. This museum expresses spontaneity, fluency of space and continuum and has no physical boundaries. It weaves a web of rationality and emotion subtly merging with one another.

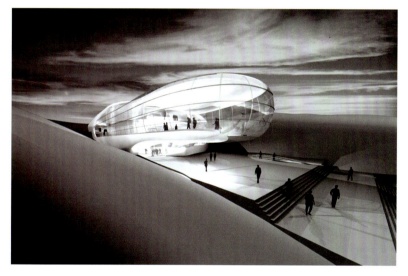

循环
circulation

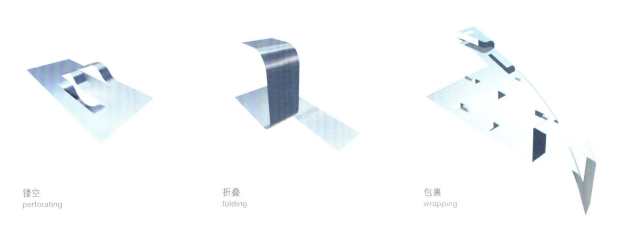

镂空
perforating

折叠
folding

包裹
wrapping

商业和贸易展览中心，比勒费尔德，2000年
铁轨旁的一只银色翅膀

每个城市沿着铁路线、高速公路或河岸旁的商店令人有些惊奇。比勒弗尔德市将紧挨火车站和市政厅，面积为3公顷的一区域作为开发用地。

COMMERCIAL AND TRADE FAIR CENTRE, BIELEFELD, 2000
A SILVER WING ALONGSIDE THE RAILS

Every city has a few surprises in store alongside railway lines, expressways or river banks. It is their urban potential that we need to explore. In the city of Bielefeld an area of three hectares of land alongside the railway tracks close to the railway station and the city hall was up for development. This was the

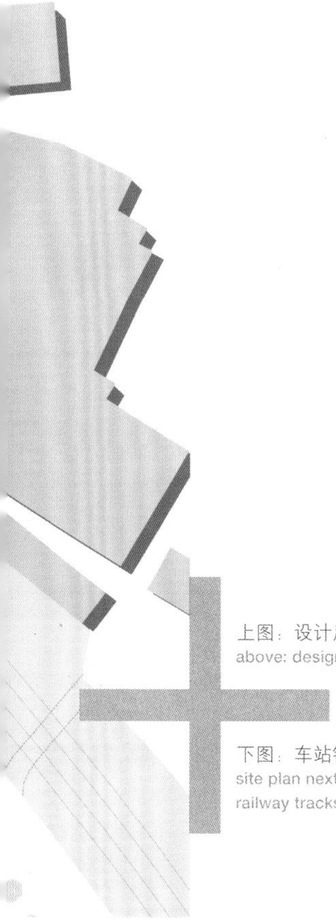

上图：设计原理
above: design principles

下图：车站铁轨旁的基地平面
site plan next to
railway tracks

选定在这个25000m²的区域范围里建一家新的旅馆、一家百货商店、一些商铺和一个贸易展览中心。

为满足大规模的计划和建筑所处环境的诸多需求，设计方案将这座建筑意欲涵盖的各种用途和活动转化成简洁的体形。所有区域被统一在一个形如银质羽翼的壳体之下，创造出功能完善和时尚美感的独特都市空间。

一座50m高的办公塔楼增加了引人注目的垂直尺度，并且构成了城市的北部入口的视觉引导。面对火车站是一个玻璃结构、一个大的陈列橱窗，在视觉上将城市与其内部的大型购物中心和旅馆联系起来。

空中花园打碎了建筑体量，并使陈列区域沐浴在日光中。实用和诗意被和谐的统一在银色羽翼之下，并在比勒费尔德市的门户位置创造了富有魅力的当代建筑。

designated site for a new hotel, a department store, shops and a fair and exhibition centre occupying an area of 25.000 square metres.

In response to this large-scale brief and the demand for a construction heterogeneous with its environment a project was developed which transforms the variety of uses and activities intended for this building into one succinct body. All areas are united under a shell in the shape of a large silver wing creating an urban space of unique character which enhances its environment in a functional and highly aesthetic fashion.

A 50 metre high office tower adds a striking vertical dimension and draws attention to the northern city entrance. Facing the railway station is a glass structure, a large display window, visually connecting the inside shopping mall and the hotel with the city. A number of patio gardens break up the volume enriching and invigorating the fair and exhibition areas with exterior spaces and daylight. Pragmatism and poetry are thus united in harmony under the silver wing providing the city of Bielefeld with an attractive contemporary building at its city gates.

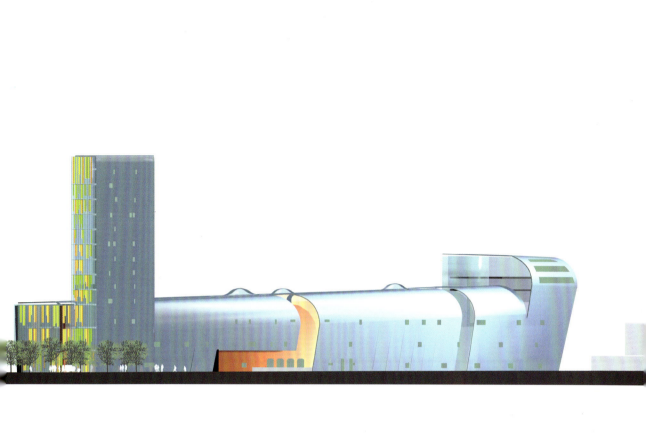
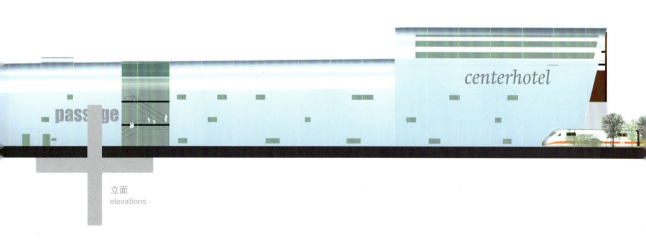

立面
elevations

体育竞技场，柏林，2005年
一个大地景观中的体育事件

在柏林外围的温斯多特正在建造一个新的大型体育场，如同一个体育活动的事件被深深地植入在大地景观中。这个"体育竞技场"是温斯多特新的健康、运动和科学公园的组成部分，可容纳12000名观众。它是一个雄心勃勃的保护和改造计划的一部分，将赋予前苏联司令部建筑新的功能和形象。

竞技场不仅能举办马术比赛，还能举办音乐会。设计构思的中心是组织联系周围的环形步道进入竞技场。

SPORTS ARENA, BERLIN, 2005
A SPORTING EVENT IN THE LANDSSCAPE

In Wünsdorf, on the periphery of Berlin, a new stadium is being constructed embedded in the surrounding landscape as a direct experience of movement. This "Sports Arena" is a building constituent of the new Health, Sports and Science Park Wünsdorf and can hold up to 12.000 spectators. It is part of an ambitious conversion project aiming to give the former Soviet Headquarters a new function and identity.

The arena will host riding events as well as concerts. A pathway running around and into the arena is the central

穿过建筑的环状运动
circular movement through the building

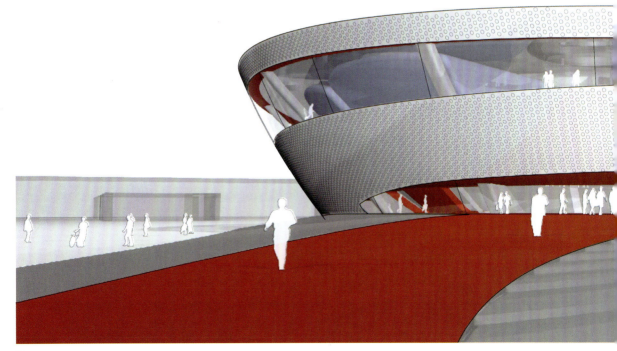

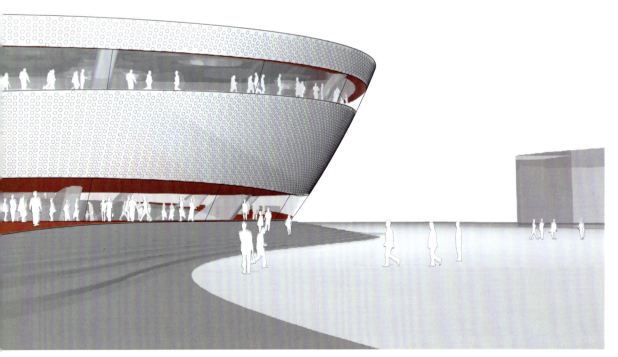

观众从下部随着自然地形进入竞技场。环形的运动趋势延伸到建筑内部。一个螺旋状通道连接外壳与大厅,赋予了建筑特有的形态。观众始终与外部自然环境保持接触。

这个碗状竞技场的特点还有最大量的自然光和更多的观看透视角度。看起来显得轻盈的木制桁架、大模数尺度的采光屋顶将跨越流线型的凹状大厅。木质和混凝土、格子轨道和圆形电镀铝板的立面将给建筑增加一种与众不同的视觉魅

design idea. The visitor approaches from slightly below and follows the natural topographical course. This movement is continued well inside where a spiral-like pathway between the outer shell and the hall endows the construction with its characteristic shape. The visitor is in constant contact with the exterior.

A maximum amount of natural light and the many viewing angles and perspectives are further remarkable features of this bowl-shaped arena. A seemingly weightless, large modular skylight roof made of wooden trusses will span the dynamic concave hall. Wood and concrete, a tartan track and a façade made of round anodised aluminium shingles will add a different visual charm to the

下图:活动空间
below: the event space

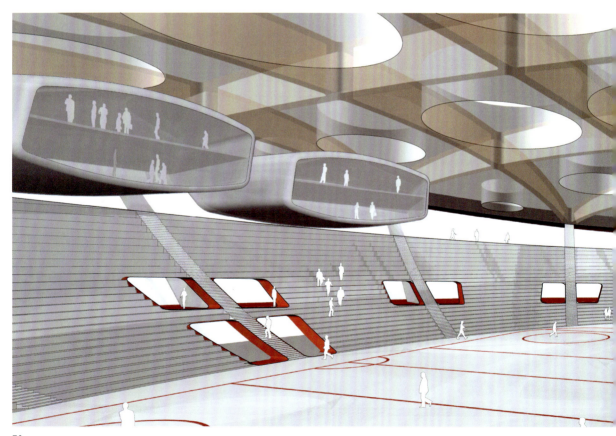

右上图：总平面
site plan

力。这种外形将使体育馆在其环境内显得动态十足。

building. It is this look that will mark this arena as a particularly dynamic event within its environment.

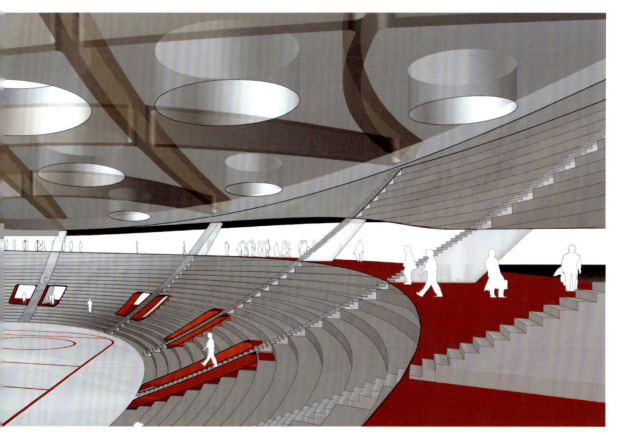

77

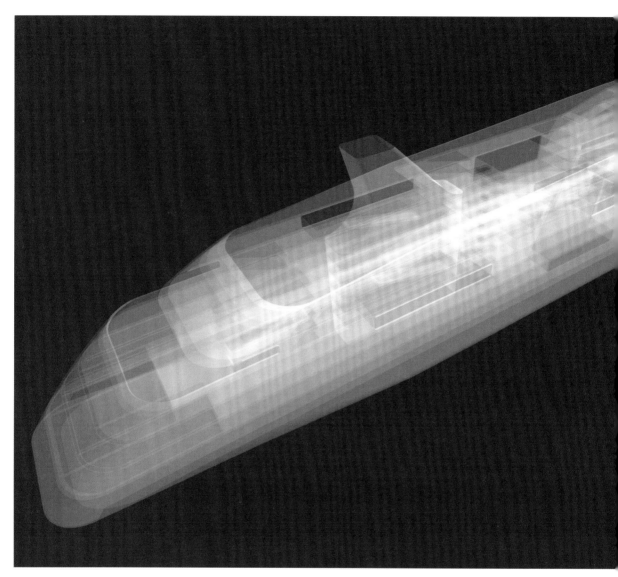
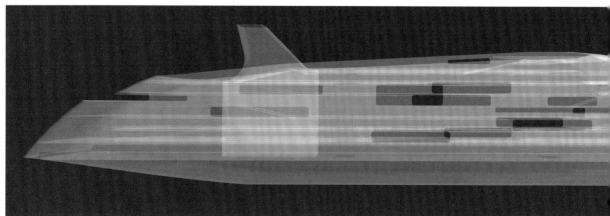

设计资料
corpus

悠闲巡游，2006 年
感官历险记

　　这艘当代的邮轮是为船主希腊人 Stelios Haji-Ioannou 和他追求的"悠闲"的服务帝国设计的，可用来满足当代年轻的游客不同的期待和需求。它被构思成一种新型可漂浮的旅馆，大大去除了以往邮轮的陈腐设计以及大多数游船老套的外观。

　　"悠闲巡游"是一种感官上的冒险，向人们提供一种新颖的方式去体验新的领域。考虑到经济的原因，对内部循环系统进行了大量的重新调整。

EASY CRUISE, 2006
A SENSUAL ADVENTURE

A contemporary cruise ship was developed for the Greek ship owner Stelios Haji-Ioannou and his "Easy" Service Empire to meet the different expectations and demands of contemporary young cruise tourists. It is conceived as a new kind of swimming hotel, far removed from the design clichés of former cruise ships or the banal appearance of mass ferries.

"Easy Cruise" is intended as a sensual adventure, a novel way for people to venture into new territories. The internal circulation system is undergoing a substantial reprogramming process follow-

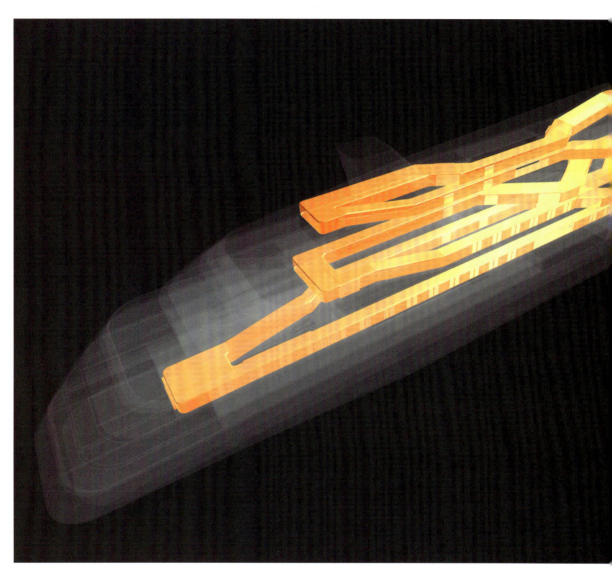
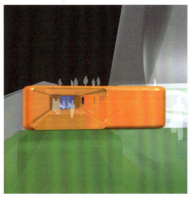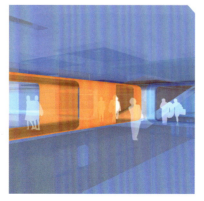

左图：木板步行道
left: boardwalk

下图：特别的去处
below: special places

通过将以前单调的走廊改造得丰富多彩，活跃的空间使它外形吸引力被大大加强。公司引人注目的"悠闲"色彩概念支配着植入新功能房间和服务区域的色彩设计。传统的"木板步行道"将会变得更加具有动感，更加丰富多彩。

如今的游客都是做好充分准备、组织性很强的个体。因此，邮轮内部和甲板上的空间，有限的空间和广阔无垠的海平面之间的空间是分别按需设计的。在这里，传统的阳光甲板、酒吧和商店引申出新意，同时互联网设备、游戏厅、俱乐部和自动投币洗衣店也增加到房间的计划中，多种功能相互交错。建筑设计运用移植和开洞等手法，以一种有趣的方式重置和延伸了舰艇上的狭窄空间。

ing economical considerations. It will be much enhanced by developing its formerly banal-looking corridors into varied, active areas. "Easy's" striking company colour will dominate the colour scheme for the new room implants and service areas. The traditional "boardwalk" will become much more dynamic and varied.

Today's clients are well-prepared and highly organised individuals. The spaces inside and on deck, the spaces between the limited world on board and the wide horizon of the sea are designed accordingly. There are new varieties of the traditional sundecks, bars and shops whereas internet facilities, gaming, clubbing areas and a launderette are added to the room scheme. Functions overlap, the architectural design is strewn with implants and perforations all serving the strategy to re-programme and extend the narrow space on board in an entertaining way.

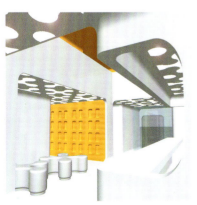

RELEASED 精选 i5+10

摄影：Claus Graubner

富士通机器人总部，斯图加特，2005年
品牌建筑随力而动

HEADQUARTERS FOR FANUC ROBOTICS, STUTTGART, 2005
BRAND ARCHITECTURE FOLLOWING THE FLOW OF FORCES

富士通项目是按照国际标准设计的。这家日本公司是自动化领域的全球领头羊。他们认为坐落在斯图加特市毗邻机场的新德国行政管理总部应该不仅仅是一个概念。为了满足展示国际形象的需求，我们采用了一个智能化系统来解决问题。这种概念是根据他们的商业哲学理念"持久运动"，同时考虑到依据当地国家津贴需求标准的不同空间使用来定位的。公司的形象被解读为：现代、功能和开放，这些特点都被精确有效地转化成为建筑语言。

The project of FANUC Robotics was designed to meet global standards. The Japanese enterprise, worldwide leaders in the field of automation, asked for more than just a concept for their new German Administration Headquarters located in Stuttgart, right next to the airport. In order to comply with the demand for global business representation an intelligent systems solution was required. The concept was to follow their business philosophy of "perpetual motion" and at the same time allow for variations in scale according to local requirements depending on national subsidy. The company's Corporate Identity was described as modern, functional and open, and these terms were indeed transferred to the architectural design with precision and efficiency.

前页图：功能主义观念
previous page: emotional functionalism

上图：穿过大厅的剖面图
above: section through the hall

下图：西立面
below: western elevation

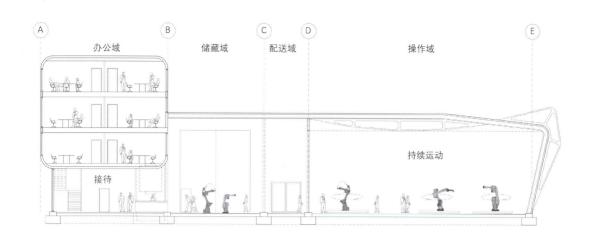

摄影：Claus Graubner

这座实验建筑设计方案基于三条高度和长度不同的平行线，从南向北延伸。

轻型无柱的东"平行线"设置剖面固定的和可变的结构。不同楼层的储藏室、培训室和行政管理区被安排在西"平行线"。两条"平行线"之间，有一条服务"平行线"，就像一根脊柱，提供所有的技术功能，同时连接着各个区域。这座建筑完全按照工业样式进行设计和建造。

由于建造采用了预制构件和优化流程的方式，整个项目从委托到交付只用了15个月的时间。

闪闪发光的银色铝壳配上大面积的玻璃和公司使用的黄色标志，使建筑在很远的地方就令人过目不忘。它的内部空间作为展示陈列室来使用，大厅的形状和顶部悬挂的钢桁架都是直接根据受力原理设计的。运用尽量少的材料真是一种非常完美的解决方案，就像制造机器人，它的任何一部分都是必不可少而又实用的。

结构与表皮的完美结合
synergy between structure and skin

预制件的精确安装
precision and prefabrication

入口区
entrance area

The building diagram of this pilot scheme is based on the idea of three "parallel lines" of variable height and length running from south to north.

The light-weight column-free eastern line accommodates demonstration and modification sections. Multi-level store room facilities, training and administration areas are to be found on the western line. Between the two, there is a service line, just like a spine, providing all technical facilities and connecting the various areas. The building was planned and constructed in a truly industrial manner.

Due to prefabrication and process optimisation it took only 15 months from the first phone call to the turn of the key.

With its smooth shimmering silver aluminium shell, large glass areas and the appliance of yellow as the company's colour, the building makes for a striking effect visible from afar. Its interior is used as a representative showroom. The shape of the hall and the minimised suspended steel trusses follow directly the laws of the distribution of force. It's an elegant solution using only a minimum of materials, just like a robot whose parts are all vital and functional.

摄影：Claus Graubner

国际航空和空间技术中心，
柏林，2003 年
流线型建筑

　　位于维尔道的国际航空和空间技术中心被设计成一个单纯的空气动力学体形。这座建筑坐落在柏林舍纳费尔德机场附近，包含了最先进的科学技术。劳斯莱斯在这里通过模拟所有可能的飞机飞行状况和高度条件，用高科技方法开发和检测他们的飞机发动机。另外，维尔道科技学院的学生也在这里进行实习。所有的器械和它多样的复合装置全部被覆盖在一个银色的流线型铝质壳体下面。

主外观
main elevation

摄影：Eberle & Eisfeld | Berlin

INTERNATIONAL CENTRE FOR AVIATION
AND SPACE TECHNOLOGY
BERLIN, 2003
„STREAMLINE ARCHITECTURE"

The "Centre for Aviation and Space Technology" in Wildau was designed as one single aerodynamic body. Located near Schönefeld Airport, Berlin, the building contains technology of the highest standard. It is here that Rolls Royce develop and test their aircraft engines in high-tech test stands simulating all possible flight and height conditions. In addition, students of Wildau Technical College may also gain practical experience here. All appliances and its various complex installations are amalgamated under a silver streamline aluminium shell.

右图:扭动的形体
left: curved shapes

摄影:Ralph Baiker

由于嵌入玻璃结构,光线被深入引导到办公空间的内部,建筑不再是一个密不透气的庞然大物,而是代表着对运动和交流的特别关注。未被开发利用的工业用地因此被重新注入了新的活力,同时通过建筑后面三个巨大的涡轮机大厅的排气塔创造了独特而强烈的视觉感受。

这种工业化设计理念不是来自飞机制造科技本身,而是来自一台符合人体工程学原理的CD播放机形象。它的结构缓缓地融合在古老的工业化环境中,同时逐渐升起一直延伸到联系柏林和德累斯顿的轻轨线。在这座建筑的中心隐藏着一个小型的花园,为全体员工提供一个绿色平静的娱乐休闲场所。

本地的、地区之间的和跨国度的活动和谐地交织,融合为空间统一体。

细部
details

  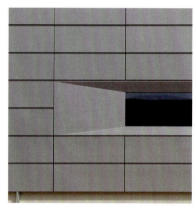

Thanks to inserted glass structures guiding the light deep into the office levels the building is far from being a hermetic monolith, but rather represents a lively focus of movement and communication. Unexploited industrial land has thus successfully been revitalised and has gained unique visible prominence by three mighty turbine hall exhaust towers at the back of the building.

The industrial design idea has been developed from the image of an ergonomically shaped CD player and not from images of aviation technology itself. The structure smoothly merges into its old industrial environment and gradually rises up to the adjacent light railway line linking Berlin with Dresden. In the centre of the building a small hidden park provides a green and calm place for the recreation of all staff.

Local, interregional and transnational movements meet in harmony and merge into a corporeal and spatial continuum.

马克西米利安和宫廷马厩广场的州歌剧院，慕尼黑，2003 年
城市的再生

STATE OPERA AND MAXIMILIANHÖFE ON MARSTALLPLATZ, MUNICH, 2003
URBAN REGENERATION

新建筑综合体位于慕尼黑市中心地带，它代表典型的马克西米利安（Maximilian）大街持续发展的独一无二的文化和经济的进步。巴伐利亚州歌剧院和两座带商店和咖啡厅的办公楼的扩建，三座建筑形成了一个新的文化广场，在这个具有历史意义的中心地带形成了一个充满吸引力的区域。过去时代的痕迹和历史建筑的残余小心翼翼地和新的建筑整合一体，像一个被偶然发现的物体暴露在规划设计和新的建筑群中。

设计的起点是创建一个原汁原

This new building compound right in the heart of the city of Munich stands for the progressive continuation of the exclusive culture and economy typical of Maximilian Street. A new cultural square defined by three new buildings, the extension of the Bavarian State Opera and two office buildings with shops and cafés, form an attractive new city quarter in the historic centre. Traces from past times and remnants of historical buildings have been carefully integrated into the new buildings and are exposed in both the urban design and the new building structure like objects found by chance.

The starting point for the design was the idea of a truly European city, a compact and complex urban grain with squares, passage ways and small streets, where

左图：总体规划图
航空图
left: Marstallplatz
aerial view

右图：排练舞台的南立面
right: south façade of
the rehearsal stage

94

摄影：Florian Holzherr

味的欧洲城市，一个有广场、巷道和小街的紧凑复合的城市文脉。城市性不是在建筑体块的内部，而是在外部的市场、街道和广场上。

公共空间和私人空间在宫廷马厩广场被有效地结合在一起。这种结合原本在战后几乎不被考虑规划发展，并因此从人们脑海中消失。

第一步，建筑师比尔克勒把马克西米利安边上的建筑内部和外墙彻底拆毁，重建成具有现代气息的零售商店和办公楼。面向马克西米利安的新的玻璃结构显示了新时代特征的改造，并与其他新建筑相联系。第二栋办公楼是方形的，与由柱子撑起的历史厅堂结合得非常完美。整个建筑体沿着一条略微扭转的中轴线建设，适应整个区域各种不同功能的公共空间和私人空间的需求，如商店、餐厅和办公楼。

这两栋建筑为巴伐利亚州歌剧院新的排演场构建了一幅美丽的背景画面。排演场有四个大型排演舞台、工作室和管理区。所有功能都被紧凑地组织在一个由玻璃和金属制造的独立的巨大体块中。闪闪发光的外壳部分是透明的，部分是不透明的，与毗邻的由列奥·冯·克伦策设计的历史建筑 Marstall 大楼分庭抗礼。宫廷马厩广场作为新的设计项目中必不可少的一部分，曾经一次又一次成为城市许多重大事件和集会的举办地。

摄影：Bilderdienst Süddeutscher Verlag

工作空间　地下室交响乐排演厅
workspaces　orchestra rehearsal hall in the basement

urbanity is manifested not inside the building mass but outside, in piazzas, streets and squares. Both public and private spaces have been carefully interlinked on Marstallplatz, a square that had scarcely been considered for urban development after the war and, as a result, had slipped to the back of people's minds.

In a first step, the 19th century building alongside Maximilianstraße by the architect „Bürklein" was gutted and stripped right down to its street façade and then rebuilt as a modern retail and office building. The new glass structure facing Marstallplatz reflects the transformation into our times and puts this building in relation to all the other new edifices. The second office building is square-shaped, elegantly integrating a historic hall of columns. Built around a slightly twisted central axis the building ensemble accommodates several diverse attractive public and private uses with shops, restaurants and offices.

Both houses set a frame for the Bavarian State Opera's new rehearsal venue, which features four big rehearsal stages, workshops and administration areas, all compactly organised in a monolith-like volume predominantly made of glass and metal. Its shiny body is partly translucent, partly opaque and represents a striking counterpart to the historic Marstall building by Leo von Klenze just next door. As an integral part of the project the newly designed Marstallplatz has once again become a place to experience, a new venue staging the city's many events and functions.

摄影：Claus Graubner   摄影：Bilderdienst Süddeutscher Verlag

一层歌剧排练厅
theatre rehearsal hall on first floor

合唱排练厅
chorus rehearsal hall

摄影: Florian Holzherr

新公共空间
new public spaces

摄影：Florian Holzherr

摄影：Florian Holzherr

上图：售票处
中图：内部／外部
above/centre:
ticket office
inside/outside

摄影：Alberto Ferrero

下图：历史建筑柱廊厅与新建筑结合
below: integration of a historic column hall into the new building

摄影：Jens Willebrand

设计：TRIAD Projektgesellschaft with Axel Büther 和贝克尔・格瓦斯・昆・昆事务所（BGKK），项目主持人：艾克・贝克尔

"星球"夜景
night shot of the planet

M星球—贝塔斯曼展馆，2000年世博会，汉诺威，2000年
通往新媒体时代的探索航程

设计：TRIAD Projektgesellschaft with Axel Büther 和贝克尔·格瓦斯·昆·昆(BGKK)，项目主持人：艾克·贝克尔

通过这座名为"M星球"的临时展馆，贝塔斯曼传媒集团在汉诺威2000年博览会上向世界展示了自己。"M"表示"为人服务的媒体(Media)"。这成为在数字媒体时代揭示整个建筑新概念的格言——建筑内部通体柔韧，同时对我们的观感提出挑战。这个展馆设计预想一艘发光的椭圆形宇宙飞船盘旋在9m高的钢柱上。

PLANET M – BERTELSMANN PAVILION EXPO 2000, HANOVER, 2000
A VOYAGE OF EXPLORATION INTO THE AGE OF THE NEW MEDIA

Design: TRIAD Projektgesellschaft with Axel Büther and Becker Gewers Kühn + Kühn (BGKK), partner in charge for BGKK: Eike Becker

With this temporary pavilion called "Planet m" the Bertelsmann Media Group presented itself to the world at the "EXPO 2000" in Hanover. "m" stood for "media for the people" as the motto for a

completely new concept of architecture in the era of the digital media – totally flexible in itself and at the same time posing a challenge to our perception. The diagram for this pavilion envisaged a shiny oval spaceship hovering above

人们并不从平常的入口进入新媒体世界中。这是一个空间升降平台，一次能将近200人领到"星球"中。"星球"长46m，宽36m。它的特点之一"多媒体感受"体现在连接建筑另一部分的桥上。相比之下更传统布置的一翼包括三个展览标高，用木板条饰面展示古老的纸质印刷媒体。展馆灵活的空间隔断划分系统能够满足使用者诸如展览、贸易、会议等不同的空间要求。

这艘宇宙飞船的三维壳体由不锈钢网眼制成，阿鲁普和设计伙伴公司 Triad 参与了合作设计。在詹姆斯·特雷斯（James Turrels）的柔和照明装置的映衬下，宇宙飞船变成了一颗耀眼的"媒体之星"。"新媒体"的形象就这样被"星"的视觉可变性表达了出来。

设计：TRIAD Projektgesellschaft with Axel Büther 和贝克尔·格瓦斯·昆·昆事务所（BGKK），项目主持人：艾克·贝克尔

nine metre high steel pipe columns. People entered into the world of the new media not via an ordinary entrance, but via a space lift platform, which pushed up to 200 people at any one time into the "planet". The 46 by 36 metre wide oval "planet" had particular features, such as the "multimedia experience", which gave access to the second part of the building via a bridge. This more conventionally organised wing contained three exhibition levels and presented the world of the old print and paper media, expressed by a façade of wooden slats. With the help of a flexible room partitioning system the pavilion could meet the publisher's changing needs for either exhibition, merchandising or conference space.

The spaceship's three-dimensional shell made of stainless steel mesh was created in cooperation with Ove Arup and the design partner company Triad. With James Turrels light installation softly shaping the pavilion, it was turned into a shining "media star". The image of the "new media" was thus expressed by the "star's" visual changeability.

摄影：Jens Willebrand

左图：在"星球"内部
left: inside the "planet"

右图：通过空间升降机进入到"星球"
right: access into the "planet" via a space lift

103

左图：光照下五彩斑斓的走廊
left: colourful illuminated loggias

右图：塔楼立面
right: high-rise façade

摄影：Claus Graubner

德利佳华办公塔楼，法兰克福/美因河畔，2003年
全景中的办公景观

这幢建筑位于新开发区"西城区"，靠近法兰克福展览中心。它是德利佳华投资银行的新建塔楼，理性、抽象而不乏惊奇与情感。首先，设计把巨大的建筑体量化解为融入天空的建筑轮廓。超现代和高度可变的标准办公空间设计要求建筑明晰、庄重。但是一些没有考虑的空间特性，诸如休闲和交流区，被加入了更多的意义和精神价值，而不仅仅是紧凑的建筑结构。

该设计简洁清晰。两座U形的7层裙房包围着纤细的塔楼。建筑综合体入口是一个5层高中庭，就像一个栩栩如生的广场。

OFFICE TOWER FOR DRESDNER KLEINWORT WASSERSTEIN, FRANKFURT/MAIN, 2003
INSIDE PANORAMAS IN AN OFFICE LANDSCAPE

Located in the new development area "City West" near Frankfurt Exhibition Centre this new tower for Dresdner Kleinwort Wasserstein Investment Bank marries a rational, abstract office brief with surprising and emotional aspects. The task on hand was firstly to reduce the impact of the large building mass with a contour that weightlessly merges into the sky. Clarity and formality were required for the design of the ultramodern and hyperflexible standardised office world. But then unexpected features, such as rooms for relaxation and communication, were inserted adding further meaning and emotional value to a rather compact building structure.

The design diagram is simple and clear. Two u-shaped seven-storey perimeter buildings frame a slim tower. Access to the complex is gained via a five-storey atrium resembling a lively piazza. The

摄影：Claus Graubner

全景中的桌球休息室
table football lounge
as inside panorama

摄影：Soup Berlin

人们可以通过电梯和轻质吊桥到达更高的楼层。位于楼顶的一间两层休息室提供了观看法兰克福全景的绝好机会。

通过与柏林艺术家团体"Soup"的合作为休闲而设计了新的休息室类型。作为休息室或者平台，色彩斑斓的廊空间被从裙房中划分出来。这些"内在的风景"是易于理解的三维形象，与严谨的办公室空间设计形成鲜明对比。另一个与整体设计概念有关的特点是吊顶的灯光设计。这种隐藏结构的例子运用多种颜色的像素屏幕和镜子，拓展了空间的广度和人对空间的观感。

upper levels are reached via lifts and light-weight bridges, a two-storey sky lounge at the very top allows for an impressive panoramic view across Frankfurt.

In cooperation with the Berlin artists' group "soup" new types of rooms for relaxation have been developed. Colourful loggia areas have been cut out of the perimeter buildings as either lounges or terraces. These "Inside Panoramas" are three-dimensional accessible images representing a clear contrast to the strict design of the office areas. Another feature integral to the overall concept is the lighting design of the casino ceiling. With the appliance of multicoloured pixel screens and mirrors this fine example of "merging structures" visually widens the space itself as well as our perception of it.

摄影：Claus Graubner

左下图：中庭
below left: atrium

右下图：隐藏结构的吊顶
below right:
casino ceiling with
"merging structures"

摄影：Ralph Baiker

107

摄影：Jens Willebrand

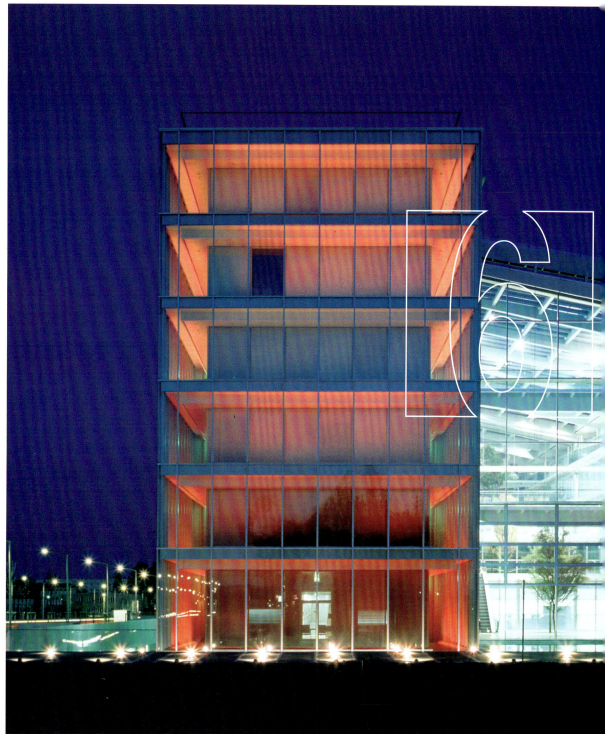

设计：贝克尔·格瓦斯·昆·昆　项目主持人：艾克·贝克尔

能源引导设计
energy as design guide

摄影：Jens Willebrand

设计：贝克尔·格瓦斯·昆·昆 项目主持人：艾克·贝克尔

看与被看——中庭
to see and to be seen
– the atrium

联网天然气股份公司总部，莱比锡，1997年
共生建筑

设计：贝克尔·格瓦斯·昆·昆
项目主持人：艾克·贝克尔

在莱比锡市的边缘将建造联网天然气股份公司新的管理总部大楼。这个项目设想在组织和技术部门之间建立协同关系，并且赋予城市边缘一块不起眼的区域以新生活的质量。

位于公司心脏部位的大型数据分理中心是我们设计概念的出发点。数据中心和较少使用资源房间一并设计，但是它的重要地位丝毫没有被减弱。其共生建筑的形象由数据进程中心发展而来，数据进程中心就像一块"热石"使中庭及两翼的房间充满了温暖。

在炎热夏季的白天，由两个体块加热系统产生的能量能够被吸收然后再进入到建筑制冷系统中。

室内规划中心是装在中庭的"热石"，一个被塑造成多用途的裂缝空间。多层绿色平台提供了每日交流和塑造文化景观的空间。所有的区域都被一个玻璃壳保护，可以对气候的变化产生敏感的反应。他们和一个照明系统联系起来，著名艺术家詹姆斯·特里尔的雕塑展示了色彩的变化从淡粉色到橘色、红色、绿色、青绿色、蓝色，投射到邻近会议区域的墙上。引人注目的视觉效果实质上提高了建筑周围区域的品质。

HEADQUARTERS FOR VERBUNDNETZ GAS AG, LEIPZIG, 1997
A SYMBIOTIC BUILDING

Design: Becker Gewers Kühn + Kühn
Partner in charge: Eike Becker

On the periphery of the city of Leipzig, the new administration headquarters of the Verbundnetz Gas AG, one of Germany's largest energy supply companies, was developed. The intention for this project was to create synergies between the organisation and technology departments and to endow an unassuming city district on the periphery with a new quality of life.

The large data processing centre at the heart of the company was the point of departure for our design concept. It was to be integrated into the room scheme sparingly using ressources, but nevertheless clearly demonstrating its significance. The image of a symbiotic building evolved from it, a "warm stone", the data processing centre, filling its environment, the atrium and the two adjacent office wings, with warmth.

摄影：Jens Willebrand

On a hot summer's day, however, the energy produced locally by two block heating systems can be absorbed and then re-directed into the building's cooling system.

Central to the room scheme is the "warm stone" embedded in the atrium, a gap space fashioned into an area of multiple use, with green terraces frequented for day-to-day communication and cultural events alike. All areas are covered by a protective glass shell whose differentiated sensors react to climatical changes. They are linked to a lighting installation, a sculpture by renowned artists James Turrell, which shows changes in colour from pale pink to orange, red, green, turquoise and blue projected onto the walls adjacent to the conference areas. This visual attraction essentially enhances the area surrounding the building.

左图：詹姆斯·特雷尔设计的灯光装置
light installation by James Turrell

右图：内部与外部之间
between inside and outside

蓝色时光
blue hour

戴姆勒克莱斯勒航空大楼，柏林，2000年
形式追随技术

设计：贝克尔·格瓦斯·昆·昆
项目主持人：艾克·贝克尔
完成：格瓦斯·昆·昆

戴姆勒克莱斯勒航空公司在柏林附近路德维希斯费尔德新建的客户服务和培训中心呈现一种"开放的建筑"，展示出一个公司面向未来的理念和技术。基地建筑物正入口处为行政办公室、涡轮引擎的维护和客户训练设备。这里需要技术的表现，组织灵活高效的工作程序。

模仿涡轮引擎中空气蒸发、压缩和加速的动作，我们提出一个建筑设计概念，强调房间之间流畅的连续性，提供多样性的体验和近距离的交流。紧凑的房子中间是一个椭圆形的中庭，围绕着开放的回廊。设计希望探索空间的运动性，并在集中交流区和私密安静区提供办公室。

摄影 Jens Willebrand

设计：贝克尔·格瓦斯·昆+昆　项目主持人：艾克·贝克尔

DAIMLERCHRYSLER AEROSPACE, BERLIN, 2000
FORM FOLLOWS TECHNOLOGY

Design: Becker Gewers Kühn + Kühn
Partner in charge: Georg Gewers
Completion: Gewers Kühn + Kühn

The new customer services and training centre of DaimlerChrysler Aerospace in Ludwigsfelde near Berlin is presented as an "open house" displaying a company philosophy and technology geared towards the future. The building at the very entrance of the site accommodates administration offices and customer training facilities for the maintenance of turbine engines. Here, technological representation and a flexible, efficient organisation of the work processes were called for.

Modelled on the action of a turbine engine, its air streams, condensations and accelerations, an architectural concept has been developed which emphasises a fluent continuum from room to room offering a variety of experiences and exchanges in close physical proximity. The compact house is developed around an oval atrium with open galleries. It explicitly wants to be explored in movement and offers flexible office areas for intensive communication as well as more reclusive, quiet areas.

工业精密度	公司的中心	航空学设计指南
industrial precision	centre of the company	aviation as design guide

在入口，一栋流线型的建筑带着三个钢片就好像三个翅膀，但是事实上是为了在双层玻璃上面形成阴影。一种叫做"Structuran"的建筑材料第一次在这座建筑北立面的商务厅使用。这种半透明并且高度保温隔热的建筑材料来自回收电视的显像管，使建筑像一块"玻璃石"。在这里我们赋予建筑独特的品质应对每一个挑战。

摄影：Jens Willebrand

设计：贝克尔·格瓦斯·昆·昆
项目主持人：艾克·贝克尔

At the entrance the house swings back like an aerodynamic volume with three sketch-like steel bodies, which may be regarded as wings, but actually provide shade for the double-shell glass façade. For the "cold" north façade cladding the commercial halls a new material called "Structuran" has been applied for the first time. This translucent and highly-weather resistant material is won from recycled TV cathode ray tubes. It endows the house with the look of a "glass stone". So each and every challenge has been met with a solution of unique quality.

设计：贝克尔·格瓦斯·昆·昆　项目主持人：艾克·贝克尔

摄影：Jens Willebrand

左图：旅行的开始和结束
beginning and end
of a journey

新火车站，东柏林，2000年
进入首都旅程的开始与结束

设计：贝克尔 · 格瓦斯 · 昆 · 昆
项目主持人：艾克 · 贝克尔
完成：格瓦斯 · 昆 · 昆

火车站是个特殊的地方，它是人们彼此告别而又能再次相见的地方，是我们在流动的世界中热切会面的地方。然而，许多火车站包括前东柏林Ostbahnhof站，冷冰冰的拒人于千里之外，缺乏必要的导向感。我们的任务是用真正的21世纪的风格使其现代化。这个曾经在原德意志民主共和国时期被改建的建筑曾经是黑暗、不透光并且杂乱无章的。在不妨碍车站服务的前提下，我们仅保留其支撑结构，并且把火车站改变成一个为城市人流服务的明亮的地方。

我们的意图是使其转变成一个拥有清晰结构、明亮光线并且亲切宜人的火车站，使其转变成一幢符合城市和经济标准的当代公共交通建筑。

NEW RAILWAY STATION, BERLIN EAST, 2000
BEGINNING AND END OF A JOURNEY INTO THE CAPITAL

Design: Becker Gewers Kühn+Kühn
Partner in charge: Oliver Kühn
Completion: Gewers Kühn+Kühn

Railway stations are special places, places where people bid each other farewell and where they meet again, places of intense personal contact in our mobile world. However, many railway stations, including the former "Ostbahnhof" in East Berlin, can appear defyingly cold and disorientating. Our task was to modernise it in true 21st century style. The building, which had been converted during the GDR regime, used to be dark, opaque and confusing. Without interrupting the train services at this station we had it gutted right down to its supporting structures and transformed it into a bright place of urban mobility.

Our intention was to turn it into a clearly structured, light and visitor-friendly railway station, into a contemporary building for public transport meeting urban and economical criteria alike. The aim was to facilitate orientation within the station

设计目标是在车站大厅里具备方向感并且重新发现自然光线的诗意导向。这座火车站不仅是对过去铁路运输时代的回忆,更应该成为时代变迁和进步的标志。

入口大厅的特点是拥有一个宽阔而又透明的悬臂屋顶。一个全玻璃的正立面显示出火车站向城市开放的姿态,玻璃用作大型商店展示的橱窗。自然光线被引入建筑内部,建筑的入口大厅中有近2000m² 的商店和办公室。钢和玻璃构成的轻质屋顶洒下柔和的自然光线,自动扶梯、艺术画廊和信息亭沐浴其中,显得生气勃勃,柔和地沐浴在日光中。这样,"Ostbahnhof"车站被融合到城市中,成为来到这个城市的旅游者印象深刻的第一幕剧情。

concourse and to rediscover the poetic dimension of natural light. Recalling the pioneering times of railway transport, this station should once more become a symbol of change and progress.

The generous entrance area is marked by a wide transparent cantilevered roof. An all-glass façade demonstratively opens the station up towards the city in the manner of a huge shop display window. It allows for natural light to be guided well inside the interior where a number of modern shops and offices, built in an area of approximately 20.000 square metres, frames the entrance hall. Escalators, galleries and information pavilions are dynamically displayed, softly bathed in filtered daylight which shines through a seemingly weightless roof made of steel and glass. Thus, the "Ostbahnhof" has been re-integrated into the city as a visually attractive opening episode to a journey into the metropolis.

鸟瞰图
aerial view

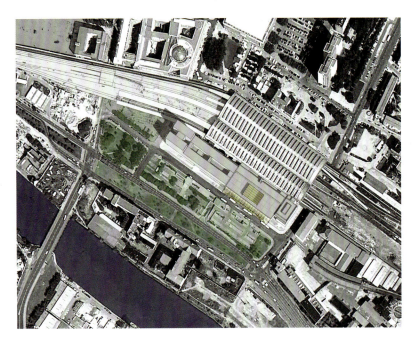

日光和透明
daylight and transparency

设计：贝克尔·格瓦斯·昆·昆　项目主持人：艾克·贝克尔

摄影：Jens Willebrand

音乐会和娱乐建筑
Admiralspalast，柏林，2007年
多彩多姿的城市

该 Admiralspalast 位于德国首都市中心，是柏林第一个大型私人出资建设的演唱会及娱乐建筑。

一个位于"黄金20世纪"充满活力的吸引人的地方，废弃数十年，它现在是一个最被经常光顾的文化场所。

重建的概念是一种多功能使用

CONCERT AND ENTERTAINMENT BUILDING
ADMIRALSPALAST, BERLIN, 2007
SWINGING CITY

The Admiralspalast in the city center of Berlin is the first large concert and entertainment building with private funding in the German capital.

A vibrant magnet in the "golden twenties" and derelict over decades it is now one of the most frequented cultural locations.

The redevelopment concept was for a multifunctional use with a large concert hall

的地方，拥有2000个座位的一个大型音乐厅、剧院以及一个大型的舞蹈俱乐部，在地下层布置安康区域和高贵豪华的咖啡厅。

设计的理念很简单，尊重现有的建筑环境：在原有环境的高品质的基础上，拿走它多余的部分并添加它所需要的部分。

一部新的雕塑楼梯将各楼层联系在一起，并让参观者亲身体验建筑的各种用途。

在古典音乐会之后，人们能在酒吧会面，然后一起去俱乐部。或在咖啡厅开始一顿早餐，如果他们喜欢还可以去安康区或者在爵士乐俱乐部会面。

在2006年8月，新的admiral-spalast重新开放之后，其重新成为了柏林最受欢迎的聚会场所之一，每天都吸引越来越多的游客。

with some 2.000 seats, a theatre as well as for a large dance club, wellness areas and the noble Grand Café at ground floor.

The design philosophy is simple and respects the listed existing building stock: it preserves where it has great quality, it takes out what is superfluous and it adds whatever is needed.

A great new sculptural staircase links all levels together and allows the visitors to experience the building with its various uses.

After a classic concert the people can meet in the bar and go to the club later. Or the start with a breakfast in the Café, go to wellness temple and meet in the Jazz Club-just as they like.

After its reopening in August 2006 the new Admiralspalast is once again one of the favourite meeting places in Berlin, attracting more and more visitors every day.

"SOPHIE-GIPS-HÖFE",柏林,1997年
一个艺术和都市文化的场所

设计:贝克尔·格瓦斯·昆·昆
项目主持人:艾克·贝克尔

把一个以前的工厂场址改造成一栋城市建筑的组成部分是柏林"Sophie-Gips-Hofe"设计的任务。遵循创新的使用理念在柏林市米特区域的中心创造了新的公共空间。它把新老建筑紧密联系在一起,而且把城市的职业生活和私生活紧密联系在一起。在这里,一个2500m²宽阔的屋顶画廊用以存放艺术家Erika和Rolf Hartmann夫妇的重要现代艺术收藏品。与此同时,一个新的走廊贯穿整个建筑,并容纳画廊、

左图：屋顶空间　roof scape
右图：入口　entrance

摄影：Jens Willebrand

设计：贝克尔·格瓦斯·昆·昆
项目主持人：艾克·贝克尔

"SOPHIE-GIPS-HÖFE", BERLIN, 1997
A PLACE OF ART AND URBAN CULTURE

Design: Becker Gewers Kühn + Kühn
Partner in charge: Swantje Kühn, Oliver Kühn

Transforming a former factory site into a building constituent in the city was the design brief for the Sophie-Gips-Höfe in Berlin. Following an innovative concept of use a new public space has been created in the very heart of the new city district of Berlin-Mitte, which closely links the old with the new and, particularly, urban professional life with private life. Here, a 2.500 square metre large penthouse gallery hosting the important collection of modern art owned by the artists couple Erika and Rolf Hartmann has been created as well as a new passage running through the entire block accommodating galleries, restaurants, loft apartments, media offices and a radio station.

The strategy was to preserve the historic traces of this listed building ensemble consisting of three industrial courtyards and to enrich it with the addition of contemporary structures to be embedded in the existing one in order to strengthen it. The old factory levels still bear the marks of intensive use, the iron and concrete supporting structure has been accentuated with white paint.

摄影：Jens Willebrand

餐厅、阁楼、媒体办公室和广播站。

设计的策略是完全保护这个被列出的由三个工业庭院组成的整体建筑的历史痕迹，同时通过把当代结构植入原有的结构中使之加固。老工厂各层仍然保留着以前的使用痕迹，铁和混凝土的支撑结构用白色的喷漆加以强调。

艺术收藏家的屋顶画廊是建筑屋顶的延伸，轻型钢和玻璃结构大厅延长了整个院子的长度。屋顶画廊明显地附加在老工业建筑之上，并且提供了一幅面对柏林众多屋顶的迷人画卷。拥有许多宽敞单元房间的现代建筑和院子中展示的白色雕塑面向 Gipsstrβe 的场地呈闭合状。就好像这些院子的其他区域一样，它追随着由各位艺术家展开的一个整体的概念。

游泳池　图书馆　下一页：沙龙
pool　library　next page: salon

摄影：Ludger Paffrath

设计：贝克尔·格瓦斯·昆·昆
项目主持人：斯万提·昆、奥利弗·昆

The art collectors' penthouse gallery is a roof extension, a concourse of light-weight steel and glass structures stretching the whole length of the courtyards. It is sensitively attached to the old industrial building and offers a charming panoramic view across Berlin's rooftops. A contemporary house with spacious flats and a new white sculpture displayed in its courtyard closes the site towards Gipsstraße. Like all the other areas in these courtyards it adheres to an integral concept especially developed by various artists.

设计：贝克尔·格瓦斯·昆·昆
项目主持人：斯万提·昆，奥利弗·昆

RELEASED
精选 i5+10

传记

斯万提·昆

2003 年	Höxter Lippe 应用科学大学，建筑理论与室内设计教授
2000 年	格瓦斯·昆·昆建筑师事务所，合伙人
1991 年	成立贝克尔·格瓦斯·昆·昆事务所
1989—1991 年	伦敦理查德·罗杰斯及合伙人事务所，建筑师
1989 年	毕业于德国慕尼黑工业大学
1986	非洲马拉维工程勘测中心项目
1983 年	美国伊利诺伊州卫斯理安大学，文学士
1964 年	出生于慕尼黑

奥利弗·昆

2003 年	2003 年德国房地产奖
2000 年	格瓦斯·昆·昆建筑师事务所，合伙人
1999 年	瑞士圣大学商务学士学位
1991 年	成立贝克尔·格瓦斯·昆·昆事务所
1989—1991 年	伦敦理查德·罗杰斯及合伙人事务所建筑师
1988 年	毕业于慕尼黑工业大学
1962 年	出生于雷根斯堡

格奥尔格·格瓦斯

摄影：Udo Hesse

2004 年	菲律宾热带雨林鸟类考察
2000 年	格瓦斯·昆·昆事务所，合伙人
1991 年	成立贝克尔·格瓦斯·昆·昆事务所
1990—1991 年	伦敦诺曼·福斯特事务所，建筑师
1990 年	毕业于斯图加特工业大学
1989—1990 年	巴黎贝尔维尔高等建筑学校，DAAD 奖学金
1978–1982 年	在伯纳德·格瓦斯门下学习雕塑
1962 年	生于贝沃根（Bevergern）

BIOGRAPHIES

SWANTJE KÜHN

2003	Professor for architecture theory and interior design at the University of Applied Sciences Lippe/Höxter
2000	Gewers Kühn und Kühn Architects
	Georg Gewers, Swantje Kühn, Oliver Kühn
1991	Formation of Becker Gewers Kühn + Kühn
1989 – 1991	Architect at Richard Rogers Partnership, London
1989	Diploma at the Technical University Munich
1986	Project for the GfZ in Malawi, Africa
1983	Bachelor of Arts, Wesleyan Univ., Ill. USA
1964	Born in Munich

OLIVER KÜHN

2003	German Real Estate Award 2003
2000	Gewers Kühn und Kühn Architects
	Georg Gewers, Swantje Kühn, Oliver Kühn
1999	Business Degree at University St. Gallen, Switzerland
1991	Formation of Becker Gewers Kühn + Kühn
1989 – 1991	Architect at Richard Rogers Partnership, London
1988	Diploma at the Technical University Munich
1962	Born in Regensburg

GEORG GEWERS

2004	Ornithologic expedition in the philippine rainforest
2000	Gewers Kühn und Kühn Architects
	Georg Gewers, Swantje Kühn, Oliver Kühn
1991	Formation of Becker Gewers Kühn + Kühn
1990 – 1991	Architect at Norman Foster Associates, London
1990	Diploma at University of Technology, Stuttgart
1989 – 1990	DAAD scholarship, Ecole d´Architecture Belleville, Paris
1978 – 1982	Trained as sculptor with Bernhard Gewers
1962	Born in Bevergern

团队成员

Aldo Conti
Alexander Mayrhofer
Alexander Moritz
Andreas Enge
Anett Hadlich
Arnd Manzewski
Barbara Schlungbaum
Benedikt Schiffels
Bernd Jäger
Bettina Ludwig
Bettina Rosenbach
Charly Deda
Dirk Müller
Don J. Lee
Frederic Treheux
Georg Gewers
Guido Scheven
Henning Hesse
Jan Blaurock
Jeanette Retzlaff
Jörg-Martin Schreiber
Jürgen Stoye
Kai-Felix Dorl
Kerstin Köhler
Kristin Neise
Lee Myoung-Ju
Lennaart Sirag
Marcel Bilow
Markus Funke
Marlene Schwabe
Michael Leone
Michael Spieler
Monika Stache
Moritz Mai
Oliver Bormann
Oliver Kühn
Peter Kühling
Ramon Karges
Roland Schreiber
Stephan Werning
Stina Torén
Swantje Kühn
Thomas Nurna
Tilman Richter von Senfft
Tobias Kunkel
Türkan Öztürk
Ulrike Franke
Uwe Karl
Verena Kämpf
Volker Ruof

建筑奖项

2004 年
第 9 届威尼斯国际建筑双年展，举办 Thyseen Krupp 创新馆展览，波鸿市展作为联邦德国展厅"德国景观"的特色之一

2003 年
2003 年度房地产奖
获 2003 年商业地产类最佳奖
马克西米利安和宫廷马厩广场的州歌剧院

2002 年
英国皇家建筑师学会授予
RIBA 建筑奖
莱比锡 Verbundnetz Gas AG 新总部大楼
设计：贝克尔·格瓦斯·昆·昆
项目负责人：艾克·贝克尔

2001
德国住宅与不动产协会、德国建筑师协会及德国城市协会授予
2000 年高品质低成本重建奖
柏林第六、第七大街 Helene—Weigel 广场的生态住宅大楼重建项目
设计：贝克尔·格瓦斯·昆·昆
项目负责人：艾克·贝克尔

1998 年
Wustenrot 基金设计奖
柏林霍夫曼收藏项目 Sophie Gips Hofe
设计：贝克尔·格瓦斯·昆·昆
项目负责人：艾克·贝克尔

DEUBAU—Prize 1998 年
1998 年全国最佳建筑师新人奖
莱比锡联网天然气股份公司 (Verbundnetz Gas AG) 新总部大楼设计
设计：贝克尔·格瓦斯·昆·昆
项目负责人：艾克·贝克尔

1997 年
德国住宅与不动产协会、德国建筑师协会及德国城市协会授予
1996 年优质低成本重建奖
柏林 Sophie—Gips—Hofe 项目
设计：贝克尔·格瓦斯·昆·昆
项目负责人：斯万提·昆，奥利弗·昆

ARCHITECTURAL AWARDS

2004

. Metamorph, 9th International Bienniale Festival of Architecture in Venice Exhibition of the Thyssen Krupp Innovation Hall, Bochum featured as part of "Germany as Landscape" in the Federal Republic of Germany pavilion

2003

. Real Estate Award 2003
Prize awarded for the Bavarian State Opera, Maximilianhöfe Project on Marstallplatz Square, Munich. Named Best Real Estate of 2003 in the category "Commercial Real Estate"

2002

. RIBA Award for Architecture awarded by the Royal Institute of British Architects, for the New Headquarters Building for Verbundnetz Gas AG, Leipzig

Design: Becker Gewers Kühn + Kühn,
Partner in charge: Eike Becker

2001

. Redevelopment Award 2000
High Quality – Acceptable Cost,
for the ecological redevelopment of a residential tower at No. 6/7 Helene-Weigel-Platz, Berlin, Marzahn district, awarded by Bundesverband Deutscher Wohnungs- und Immobilienunternehmen e.V., Bund Deutscher Architekten and the Deutsche Städtetag

Design: Becker Gewers Kühn + Kühn,
Partner in charge: Georg Gewers

1998

. Wüstenrot Foundation Design Award for the Hoffmann Art Collection – Sophie Gips Höfe, Berlin

Design: Becker Gewers Kühn + Kühn,
Partner in charge: Swantje Kühn, Oliver Kühn

. DEUBAU-Prize 1998
National Award Best New Architects for the New Headquarters for Verbundnetz Gas AG, Leipzig

Design: Becker Gewers Kühn + Kühn,
Partner in charge: Eike Becker

1997

. Redevelopment Award 96 High Quality – Acceptable Cost,
for the Sophie-Gips-Höfe, Berlin, awarded by Bundesverband Deutscher Wohnungs- und Immobilienunternehmen e.V., Bund Deutscher Architekten and the Deutsche Städtetag

Design: Becker Gewers Kühn + Kühn,
Partner in charge: Swantje Kühn, Oliver Kühn

2000年以来的竞赛获奖作品

2006年
湖北美术学院，中国武汉，一等奖
慕尼黑中央火车站新站，一等奖组
城市研究"Laim MK2"，慕尼黑，一等奖
体育竞技场，柏林，一等奖

2005年
新传媒大楼，慕尼黑，一等奖
奥迪公司变速箱及排放研究中心，英戈尔施塔特，二等奖
花样住宅——垂直宫殿，高层住宅，华沙，二等奖
Karstadt ECE 购物中心，埃森，三等奖

2004年
退休老人、妇女及青年的新家庭部，柏林（被采用）
Ebene 法兰克福／美因机场，二等奖
慕尼黑新中心火车站，一等奖组
德国富士通机器人新总部大楼，斯图加特，一等奖

2003年
2005年日本爱知县世界博览会德国展馆，二等奖
（与ART+COM and dan pearlman 合作）

2012年 "体育动感" 奥林匹克复合体，莱比锡，二等奖
西门子AG 医学科技博物馆，埃尔兰根（被采用）

2002年
Sciddeutscher Verlag 新总部德国出版公司大楼，慕尼黑，一等奖
Charlottenburg 生物科技园，柏林，一等奖
沃尔夫斯堡金属工业联盟总部新楼，二等奖
TKS 创新馆，波鸿，一等奖

2001年
奥迪品牌研究所与饭店和会议中心，英戈尔施塔特，一等奖
马克西米利安和宫廷马厩广场新巴伐利亚州歌剧院排演建筑，两个办公楼、新公共空间，慕尼黑，一等奖

2000年
水边住宅（临水而居），柏林，第一名
OWL 购物中心，比勒费尔德，第二名

PRIZE WINNING COMPETITION SINCE 2000

2006
- Hubai Fine Arts University, Wuhan/China (1st Prize)
- New Central Railway Station in Munich (1st Prize group)
- Urban Study "Laim MK2", Munich (1st Prize)
- Sports Arena, Berlin (1st Prize)

2005
- New Building for a Media Agency, Munich (1st Prize)
- Gearbox and Emissions Centre Audi Ltd, Ingolstadt (2nd Prize)
- Blossom House – Palais Vertical, New Residential High Rise, Warsaw (2nd Prize)
- Shopping Centre for Karstadt ECE, Essen (3rd Prize)

2004
- New Ministry for Families, Senior Citizens, Women and Youth, Berlin (Purchase)
- Ebene 0, Frankfurt/Main Airport (2nd Prize)
- New Central Railway Station in Munich (1st Prize group)
- New Headquarters Building for Fanuc Robotics Germany, Stuttgart (1st Prize)

2003
- German Pavilion at the Expo 2005 in Aichi/Japan (2nd Prize)
  with ART+COM and dan pearlman
- Olympic Building Compound "Sport Moves", Leipzig 2012 (2nd Prize)
- Museum of Medical Technology, for Siemens AG, Erlangen (Purchase)

2002
- New Headquarters Tower for Süddeutscher Verlag as major German publishing company, Munich (1st Prize)
- Biotech Park Charlottenburg, Berlin (1st Prize)
- New Headquarters Building for the Metal Industry Union Wolfsburg (2nd Prize)
- TKS Innovation Hall for ThyssenKrupp, Bochum (1st Prize)

2001
- AMI – Audi Brand Institute with Hotel and Conference Centre, Ingolstadt (1st Prize)
- Maximilianhöfe on Marstallplatz, with new rehearsal building for the Bavarian State Opera, two office buildings and major new public space, Munich (1st Prize)

2000
- Living by the Water, Berlin (1st Prize)
- OWL-Shopping Centre in Bielefeld (2nd Prize)

精选作品

法国富士通机器人新总部大楼，巴黎（2006 年）

新慕尼黑传媒新大楼，慕尼黑（2005 年）

Admiralspalast 剧院重建，柏林（2005 年）

德国斯图加特富士通机器人新总部大楼（2005 年）

Suddeutscher Verlag 新总部大楼，慕尼黑（2005 年）

投资银行高层建筑，法兰克福（2004 年）

马克西米利安和宫廷马厩广场州歌剧院，慕尼黑（2003 年）

Charlottenburg 生物科技园，柏林（2003 年）

波鸿 Thyssen—Krupp—Stahl 创新馆，（2003 年）

劳斯莱斯新航空航天中心，维尔道（2003 年）

AMI 奥迪品牌研究所，英戈尔施塔特（2001 年）

商业、商务服务及旅游协会大楼，柏林（2000 年）
设计：贝克尔·格瓦斯·昆·昆
项目负责人：乔治·格瓦斯

戴姆勒克莱斯勒航空大楼移动式测试装置维护，柏林－勃兰登堡（2000 年）
设计：贝克尔·格瓦斯·昆·昆
项目负责人：乔治·格瓦斯

Helene—Weigel—Platz 6／7，柏林（2000 年）
设计：贝克尔·格瓦斯·昆·昆
项目负责人：乔治·格瓦斯

2000 年世博会贝塔斯曼展馆，汉诺威（1999 年）
设计：TRIAD Projektgesellschaft with Axel Büther 和贝克尔·格瓦斯·昆·昆
项目负责人：艾克·贝克尔

Ostbahnhof 未来火车站，东柏林（1999 年）
设计：贝克尔·格瓦斯·昆·昆
项目负责人：奥利弗·昆

Sophie—Gips—Hofe，Hoffmann 收藏，柏林（1997 年）
设计：贝克尔·格瓦斯·昆·昆
项目负责人：奥利弗·昆

Verbundnetz Gas AG 新总部大楼，莱比锡（1997 年）
设计：贝克尔·格瓦斯·昆·昆
项目负责人：艾克·贝克尔

Friedrichsthal 住宅发展计划，什未林（1996 年）
设计：贝克尔·格瓦斯·昆·昆
项目负责人：奥利弗·昆

SELECTION OF BUILDINGS

- New Headquarters Building for Fanuc Robotics France, Paris (2006)

- New Building for a Media Agency, Munich (2005)

- Reconstruction of the Theatre "Admiralspalast", Berlin (2005)

- New Headquarters Building for Fanuc Robotics Germany, Stuttgart (2005)

- New Headquarters Building for Süddeutscher Verlag, Munich (2005)

- High-Rise for an Investment Bank, Frankfurt/Main (2004)

- Bavarian State Opera House, Maximilianhöfe, Marstallplatz Munich (2003)

- BioTechPark Charlottenburg, Berlin (2003)

- Thyssen-Krupp-Stahl Innovation Hall, Bochum Stahlhausen (2003)

- New Centre for Aviation and Aerospace for Rolls Royce, Wildau (2003)

- AMI-AUDI Brand Institute, Ingolstadt (2001)

- Association Building of "Commerce, Commercial Service, Tourism", Berlin (2000)
  Design: Becker Gewers Kühn + Kühn,
  Partner in charge: Georg Gewers

- DaimlerChrysler Aerospace – MTU Maintenance, Berlin-Brandenburg (2000)
  Design: Becker Gewers Kühn + Kühn,
  Partner in charge: Georg Gewers

- Helene-Weigel-Platz 6/7, Berlin (2000)
  Design: Becker Gewers Kühn + Kühn,
  Partner in charge: Georg Gewers

- Bertelsmann Pavilion EXPO 2000, Hanover (1999)
  Design: TRIAD Projektgesellschaft with Axel Büther and
  Becker Gewers Kühn + Kühn (BGKK),
  partner in charge for BGKK: Eike Becker

- Ostbahnhof – Railway Station of the Future, Berlin East (1999)
  Design: Becker Gewers Kühn + Kühn,
  Partner in charge: Oliver Kühn

- Sophie-Gips-Höfe – Hoffmann Collection, Berlin (1997)
  Design: Becker Gewers Kühn + Kühn,
  Partner in charge: Swantje Kühn, Oliver Kühn

- New Headquarters Building for Verbundnetz Gas AG, Leipzig (1997)
  Design: Becker Gewers Kühn + Kühn,
  Partner in charge: Eike Becker

- Residential Development Friedrichsthal, Schwerin (1996)
  Design: Becker Gewers Kühn + Kühn,
  Partner in charge: Oliver Kühn

参考文献
(SELECTION OF REFERENCES)

- ABB Grundbesitz
- Admiralspalast Berlin
- Audi
- Bavarian State Opera
- BioTechPark Charlottenburg Development
- BVT – Brendel von Tuder
- CBP – Cronauer Beratung Planung
- CM Immobilien
- Deutsche Bahn Station & Service
- Deutsche Biodiesel
- DG Immobilien Management
- DIL – Deutsche Immobilien Leasing
- Dreyer Brettel und Kollegen Management
- Fanuc Robotics Germany
- Fanuc Robotics France
- FOM – Future Office Management
- GE Fanuc, Switzerland
- Hanseatische Wohnungsbau
- Hochtief
- Hydropolis
- Loxydom
- Orco Group, France
- Quantum Immobilien
- Rolls-Royce Germany
- SBP – Synergy Bauprojekt
- Schering
- City of Bochum
- City of Osnabrück
- Süddeutscher Verlag
- ThyssenKrupp Stahl
- WFG – Regionale Wirtschaftsförderungsgesellschaft Dahme-Spreewald

著作权合同登记图字：01-2007-4298号

图书在版编目（CIP）数据

诗意的功能主义——德国格瓦斯·昆·昆建筑师事务所专辑/
李保峰译.—北京：中国建筑工业出版社，2007
ISBN 978-7-112-09594-0

Ⅰ.诗... Ⅱ.李... Ⅲ.建筑设计－作品集－德国－现代
Ⅳ.TU206

中国版本图书馆CIP数据核字（2007）第115663号

Copyright © 2006 Gewers Kühn+Kühn
Translation © 2007 China Architecture & Building Press
All rights reserved.

Gewers Kühn+Kühn: 15+10 Unreleased and Released

本书由德国Gewers Kühn+Kühn建筑师事务所授权翻译出版

责任编辑：程素荣
责任设计：郑秋菊
责任校对：安　东　陈晶晶

诗意的功能主义
——德国格瓦斯·昆·昆建筑师事务所专辑

李保峰　译
*
中国建筑工业出版社出版、发行（北京西郊百万庄）
各地新华书店、建筑书店经销
北京嘉泰利德公司制版
北京中科印刷有限公司印刷
*
开本：787×1092毫米　1/16　印张：9　字数：216千字
2007年11月第一版　2007年11月第一次印刷
定价：65.00元
ISBN 978-7-112-09594-0
　　　　（16258）

版权所有　翻印必究
如有印装质量问题，可寄本社退换
（邮政编码　100037）